Sensors for Stretchable Electronics in Nanotechnology

Emerging Materials and Technologies

Series Editor
Boris I. Kharissov

Recycled Ceramics in Sustainable Concrete: Properties and Performance
Kwok Wei Shah and Ghasan Fahim Huseien

Photo-Electrochemical Ammonia Synthesis: Nanocatalyst Discovery, Reactor Design, and Advanced Spectroscopy
Mohammadreza Nazemi and Mostafa A. El-Sayed

Fire-Resistant Paper: Materials, Technologies, and Applications
Ying-Jie Zhu

Sensors for Stretchable Electronics in Nanotechnology
Kaushik Pal

Polymer-Based Composites: Design, Manufacturing, and Applications
V. Arumugaprabu, R. Deepak Joel Johnson, M. Uthayakumar, and P. Sivaranjana

Nanomaterials in Bionanotechnology: Fundamentals and Applications
Ravindra Pratap Singh and Kshitij R. B. Singh

Sensors for Stretchable Electronics in Nanotechnology

Edited by

Kaushik Pal

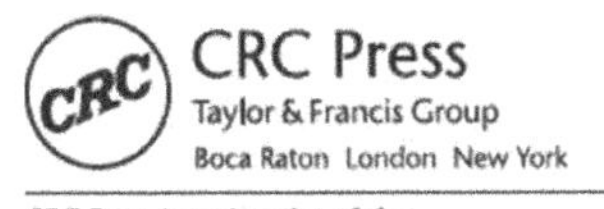

CRC Press is an imprint of the
Taylor & Francis Group, an informa business

First edition published 2022
by CRC Press
6000 Broken Sound Parkway NW, Suite 300, Boca Raton, FL 33487-2742

and by CRC Press
2 Park Square, Milton Park, Abingdon, Oxon OX14 4RN

CRC Press is an imprint of Taylor & Francis Group, LLC

Library of Congress Cataloging-in-Publication Data
A catalog record has been requested for this book

ISBN: 978-0-367-64281-5 (hbk)
ISBN: 978-0-367-64283-9 (pbk)
ISBN: 978-1-003-12378-1 (ebk)

Typeset in Times
by Newgen Publishing UK

Contents

Preface

Sensors for Stretchable Electronics in Nanotechnology explores recent trends in flexible and stretchable electronics – innovations in material synthesis, mechanical design, and fabrication strategies employing soft substrates. The biggest challenge is that the entire electronic system must allow not only bending but also stretching. Therefore, construction of stretchable conductors has become crucial for connecting working circuits of various stretchable devices that promise pathways to eco-friendly products as well as sustainable applications, potentially at industrial scale. This book comprises nine chapters from diverse fields, covering stretchable conductors. Various stretchable electronic devices – including stretchable heaters, stretchable energy conversion and storage devices, stretchable transistors, sensors, and renewable energy as well as next-generation energy storage devices – are fabricated using multiple manufacturing strategies.

In recent decades, there have been many projections on the future of stretchable materials in the next generation of smart devices; this will enable applications that are simply not possible with rigid, planar electronics. There is clearly an urgent need to further develop emerging technologies. On a different frontier, growth and manipulation of materials on the nanometer scale have progressed at a fast pace. Selected recent and significant advances in nanomaterials for nanoelectronic applications are reviewed in the book, and special emphasis is given to graphene-based materials and devices for transparent stretchable electronics (Chapter 2), elastomeric substrate for stretchable electronics (Chapter 3), highly sensitive long-term durability of wearable biomedical sensors (Chapter 4), fabrication of stretchable composite thin film for superconductor applications (Chapter 5), ultra-thin graphene assembly of liquid crystal stretchable matrix for thermal and switchable sensors (Chapter 6), CNT/graphene-assisted flexible thin-film preparation for stretchable electronics and superconductors (Chapter 7), and optical sensor-based hydrogen gas detection (Chapter 8). These accounts highlight some of the progress made and point out the factors that limit the optimization of efficiency.

Thus, it will be clear to all readers that nanotechnology-enabled flexible electronics are starting to scale up dramatically. As they become mature and cost-effective in the decades to come, they could help the move from fossil fuels and improve the performance of biofuel industry through utilization of nanocatalysts, or enable the manufacture of materials with high durability and lower weight for use in the wind energy industry. This book provides an overview of the key current developments directing future research on utilization of such tailored nanostructures or hybrid nanoassemblies, which will play an essential role in achieving the desired goal of cheap and efficient fuel and electricity.

The book is informed by evidence from academics, scientists, scholars, and engineers. It illustrates the wide-ranging interest in these areas and provides a background to the chapters which address the novel synthesis of high-yield nanomaterials and their hybrid composite matrix, graphene, polymer, liquid crystal electro-optic switching applications, nanobiotechnology, spectroscopic characterization as well as extensive electron microscopic study, flexible and transparent electrodes, optoelectronics, nanoelectronics, switchable device modulation, health care devices, energy storage devices, solar/fuel cells, and environmental and regulatory implications of various aspects of smart nanotechnologies. With appropriate regulation, commercial production of manufactured novel composite materials can be realized. Furthermore, the many brilliant discoveries and explorations highlighted throughout the book demonstrate technical excellence in high-quality inter- and cross-multidisciplinary research.

Lastly, I would like to express my overwhelming gratitude to all the contributors for their excellent research offerings in this book. I express my gratitude to the entire team at Taylor & Francis, CRC Press, particularly the Editorial Managers (Allison Shatkin, Gabrielle Vernachio), as well as the Series Editor (Kharissov Boris) for their efficient handling of this book at all stages of its publication. I am confident too that within a short interval, the book series will be popular in universities and institutes worldwide and hopefully will achieve the highest citation in coming years.

Prof. (Dr.) Kaushik Pal,
Ph.D.; D.Sc. (Malaysia); Marie-Curie (Greece);
CAS Fellow (China)
Universidade Federal do Rio de Janeiro
(LABIOS/IMA/UFRJ), Rio de Janeiro, Brazil

About the Editor

Kaushik Pal received his Doctor of Philosophy (Ph.D.) in Physics from the Government of India-recognized University of Kalyani, West Bengal, India. He also has an Honorís Causà (D.Sc.) from the Higher National Youth Skill Institute (IKTBN Sepang), Malaysia. He was awarded the prestigious Marie-Curie Postdoctoral Fellowship (Greece), offered by the European Union, and the Chinese Academy of Science (CAS) Fellowship from Wuhan University, China. As a distinguished academician, Prof. Pal has held research and teaching positions at various top-tier universities and research institutions. He was advisory Professor-in-Charge at IKTBN Sepang (Malaysia), the University of Maribör (Slovenia), and Yarmouk University (Jordan). A well-known expert supervisor and an experienced Group Leader, he is currently a distinguished Professor at Universidade Federal do Rio de Janeiro (LABIOS/IMA/UFRJ), Rio de Janeiro, Brazil. In the last academic year, Prof. Pal's research group published 110 articles in peer-reviewed (SCI/Scopus) international journals and edited/authored 25 books (published by Elsevier, Springer, Apple Academic Press, Bentham Science, NOVA, Jenny Stanford, CRC Press, and InTech publishers). Prof. Kaushik is a member of various scientific societies, organizations, and professional bodies. He supervises a significant number of bachelor's, master's, doctoral, and postdoctoral students in India and in reputed research institutions and universities overseas. Dr. Pal's research innovations include outstanding performance in green chemistry of nanomaterials, polymer, carbon nanotube/graphene/liquid crystalline optical materials, nanovaccinology, mask formulation, drug delivery and therapeutics, electro-optical polarization/electrical spectroscopy characterizations, waste-water treatment, switchable devices, and sensor nanotechnology. His work has been published by the Royal Chemical Society, Elsevier, Springer, InTech, and IEEE publications. He has published significant articles in several international top-tier journals, which are widely cited, and has authored as well as edited international book chapters and 10 review articles. He has chaired and convened almost 35 – international events, symposiums, conferences, national workshops, summer internship programs, and Ph.D. refresher courses, and he has contributed around 24 plenary presentations, 30 keynote speeches, and 35 invited lectures worldwide.

Contributors

Vinayak Adimule
Angadi Institute of Technology and Management, Savagaon Road, Belagavi, 5800321, Karnataka, India.

Alaa A. A. Aljabali
Faculty of Pharmacy, Department of Pharmaceutics and Pharmaceutical Technology, Yarmouk University, Shafiq Irshidat Street, Irbid 21163, P. O. BOX 566, Jordan.

Abu Bin Imran
Department of Chemistry, Bangladesh University of Engineering and Technology, Dhaka, Bangladesh.

Sandip Kumar Dash
Department of Zoology, Berhampur University Bhanja Bihar, Berhampur, 760007, Ganjam, Odisha.

Kamal Dua
Discipline of Pharmacy, Graduate School of Health, University of Technology, Sydney, NSW 2007, Australia.

Tahmina Foyez
Department of Pharmaceutical Sciences, North South University, Dhaka, Bangladesh.

Mohammad Harun-Ur-Rashid
Department of Chemistry, International University of Business Agriculture and Technology, Dhaka, Bangladesh.

Archita Lenka
Department of Zoology, Berhampur University Bhanja Bihar, Berhampur, 760007, Ganjam, Odisha.

Jyoti Mali
Atharva College of Engineering, University of Mumbai, India.

A. P. Meera
Research & Postgraduate Department of Chemistry, KSMDB College, Sasthamcotta, Kerala, India.

Santosh S. Nandi
Chemistry Section, Department of Engineering Science and Humanities, KLE Dr. M S Sheshgiri College of Engineering and Technology, Udyambag, Belagavi, 590008, Karnataka, India.

Kaushik Pal
Laboratório de Biopolímeros e-Sensores, Instituto de Macromoléculas, Universidade Federal do Rio de Janeiro (LABIOS/IMA/UFRJ), Centro de Tecnologia – Cidade Universitária, AV Horácio Macedo 2030, Bloco J CEP 21941-598 CP 68525, Rio de Janeiro, Brazil.

Bandita Panda
Department of Zoology, Berhampur University Bhanja Bihar, Berhampur, 760007, Ganjam, Odisha.

Narayan Panda
Department of Zoology, Berhampur University Bhanja Bihar, Berhampur, 760007, Ganjam, Odisha.

Om Prakash
MRIET, University of Hyderabad.

Sasireka Rajendran
Mepco Schlenk Engineering and Technology, Mepco Nager, Sivakasi, Tamil Nadu, 626005.

Noureddine Ramdani
Polymer Materials Research Center, College of Materials Science and Chemical Engineering, Harbin Engineering University, Harbin 150001, China.

Vinoth Rathinam
PSR Engineering College, Sivakasi, Tamil Nadu, 626140.

Chinmaya Kumar Sahu
Department of Zoology, Berhampur University Bhanja Bihar, Berhampur, 760007, Ganjam, Odisha.

Nilophar Shaikh
Research Scholar, Angadi Institute of Technology and Management, Savagaon Road, Belagavi-5800321, Karnataka, India.

P. B. Sreelekshmi
Research & Postgraduate Department of Chemistry, KSMB College, Sasthamcotta, Kerala, India.

Murtaza M. Tambuwala
SAAD Centre for Pharmacy and Diabetes, School of Pharmacy and Pharmaceutical Science, Ulster University, Coleraine, UK.

Sugumari Vallinayagam
Mepco Schlenk Engineering and Technology, Mepco Nager, Sivakasi, Tamil Nadu, 626005.

Jyothy G. Vijayan
Department of Chemistry, M S Ramaiah University of Applied Sciences, Bengaluru, India

B. C. Yallur
Department of Chemistry, M S Ramaiah Institute of Technology, Bangalore, 560054, Karnataka, India.

1 Introduction to Sensor Nanotechnology and Flexible Electronics

Alaa A. A. Aljabali, Kamal Dua, Kaushik Pal, and Murtaza M. Tambuwala

CONTENTS

1.1 INTRODUCTION

Current advances in low-power electronic smartphone and wearable devices, modern data acquisition methods, and the ICT environment have opened up prospects for a fresh approach in health and medicine. Health services are continually moving towards a preventive, patient-centered approach, replacing the old hospital-based model. In the early 2000s, developments in intelligent clothing were regarded as vital, leading to the production of fabrics with embedded sensors that connect to portable personal digital assistance systems, which minimized both morbidity and medical costs associated with the vascular system [1, 2]. Flexible electronics research has evolved dramatically over the past decade. A vast number of versatile systems like displays and sensors have been developed by researchers from around the world. Material advances have been crucial to scientific progress in this area in recent years [3]. Transistors, connectors, memory cells, passive components, and other products make it difficult to make portable electronics a reality. Nanostructures of different forms have provided a suitable medium for designing highly efficient semiconductors, dielectric materials, and conductors with nanoparticles (NPs), nanotubes, nanowires, and organic matter for various electronic medical applications [4, 5].

More recently, portable clothing devices have been considered for preventing and enhancing well-being and active aging, for successful reduction of physical age and behavioral decline, and for sustainable medical infrastructure. These objectives can be achieved through mobile and customized wellness initiatives, which pose many obstacles, such as portable electronics energy autonomy [6].

Flexible sensors can be considered as superior materials for smart sensing systems such as electronic products, prosthetics, robotics, medical treatments, protective equipment, pollution monitoring, environmental monitoring, domestic travel, and protective space travel [7]. There is an emerging trend towards producing reliable real-world flexible sensors that rely on NPs between 10 nm and 100 nm in diameter.

There are several other reasons why the use of NPs for flexible sensors (including in materials) is exciting. The first concerns the assumed capacity to synthesize almost any form of NP, if not at will then with considerable power. Several research experiments have shown that NPs are controllable, with cores made of pure metal (e.g. Au, Ag, Co, Pt, Pd, Cu, Al) or metallic composites (e.g. Pd, Pt, NiFe/Pt, Au/Ag/Cu, Au/Ag/Cu/Pd, Au/Ag, Au) [8–10]. The second reason is that NPs can be sustained with a wide range of chemical ligands: alkyl thiols and alkanaethiolates, arenethiolates, alkyl-trimethyloxysilans, xanthates, oligonucleotides, dialkyl disulfides, proteins, sugars, phospholipids, and enzymes. This suggests that NPs with a combination of chemical and physical functions can be obtained for sensing applications and can significantly impact the sensitivity and selectiveness of sensors, as shown in Figure 1.1 [11–13].

The third reason is the potential to vary the scale and form of NPs and thus the ratio between surface and volume (in spheres, rectangles, hexagons, cubes, triangles, and star-and-branch-like outlines). These features would allow for control over the surface properties of materials and the resulting 'quality,' with physicochemical characteristics like strain, temperature, plasmon resonance, and much more [14, 15].

Smartphones represent a relatively new and quickly produced commercial electronics platform that has allowed more comfortable and faster communication between people. Intensive development is now being carried out in portable electronics that can offer a much more convenient electronic interface than smartphones [16]. Importantly, many foldable electronics products/prototypes have already been introduced and can

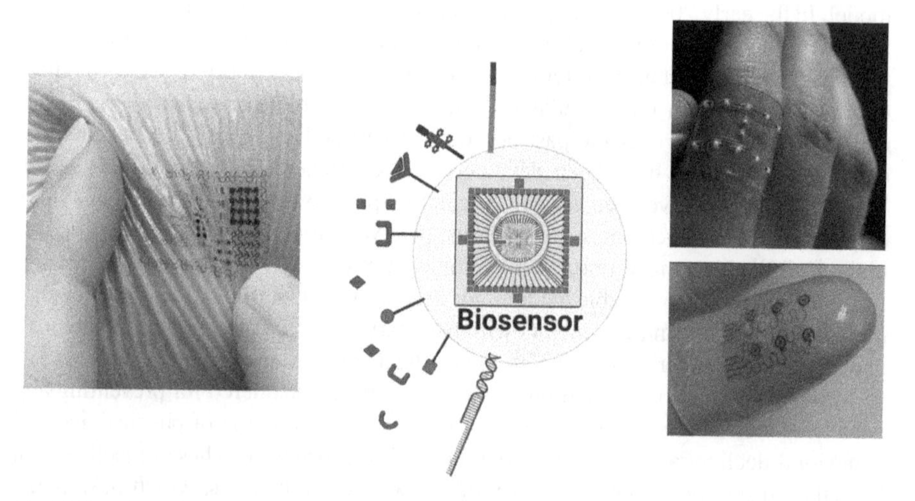

FIGURE 1.1 Schematic illustration of flexible nanoelectronics used as biosensors to monitor health and well-being.

also be categorized as wearable electronics [17, 18]. The mechanical flexibility of electronics offers an opportunity to operate on a substratum with no specified shape, and therefore mechanical versatility enables the safe movement of electronics on a wide range of object surfaces, including bodies. Thus, it is anticipated that flexible/extensible electronics can have a variety of uses. The applications provide biomedical electronics and electronics for tracking social media and the Internet and external climate monitoring systems [19–21].

Nanomaterials have superior stability in integrating versatile electronics and are suitable for diverse applications in a wide variety of surface environments. This chapter briefly presents the essential properties of nanomaterials and summarizes the advances in applying those properties in different types of instruments, such as electrical and optical devices [22].

The production of compact and stretchable electrodes is one of the most critical aspects in manufacturing flexible electronics. Different standard electrodes, including those made of gold, copper, aluminum, and indium tin oxide, have been used as conventional conductive materials because of their excellent electrical conductivity. However, because of their fragility, these are difficult to use in flexible electronics. Numerous modern materials are, therefore, being produced for use in flexible electronics. Among these new materials are 1D nanomaterials, which are currently being researched for their use as versatile conducting compounds.

A 1D nanostructure typically has metal nanostructures, metal nanomaterials, and carbon nanomaterials (CNTs). The advantages of high conductivity and excellent mechanical deformations result in structural characteristics with a high aspect ratio [23]. The 1D design provides a straight route for the transport of loads and reduces grain borders or defects. In the event of deformations, cracks are preferentially produced mostly on grain borders or other defects. Since such cracks result in a drastic rise in resistance as charging transport is removed, as much stretchability as possible is needed to avoid cracks [24]; for example, 1D nanomaterials with an area ratio above 100 nm in diameter that reflect silver nanowires (AgNWs) and iD materials with a diameter less than 100 nm that reflect cotton nanowires (CuNWs).

In addition, they have a sheet resistance smaller than 20 Ω per sq^{-1} and propagation of 85% or higher in the AgNWs' permeating networks. A large number of studies have also been done on flexible transparent electrode products to replace conventional porous TiO. The interactions between the nanowires have a significant adverse impact on these AgNWs' excellent mechanical, optical, and electrical properties. Electrons travel by a single-crystalline nanowire in an individual nanowire such that the AgNW has very low resistance; however, the electron flow in the junction is exceptionally resistant due to crystallinity malfunctions (of a few kilo-ohms approximately) [25, 26]. By putting AgNWs between two layers of polydimethylsiloxane (PDMS), all these nanostructures can be applied as highly stretchable and responsive strain sensors. Amjadi et al. [27] produced nanocomposites with increased stretching power in a sandwich structure. Various techniques have been used to weld interfaces between nanowires, including thermal rectification, mechanical strain, and plasma treatment. The thermal-sensitive polymers could vibrate when heated by traditional methods; however, Tokuno et al. [28] found that mechanical pressing to 25 MPa at an ambient temperature tightly linked the AgNW intersections. By forming AgNW junctions via light-induced plasmonic nanowelding, Garnett et al. [29] decreased the resistance by more than one factor of 1,000. Without forging in the heating phase, the layer is fragile to high temperature but is not degraded or denatured [30, 31].

The superior electromechanical properties of the gold nanowires (AuNWs) demonstrate good resistance to corrosion, biocompatibility, and tolerance to low contact between AgNWs and CuNWs with p-type semiconductor materials. Many experiments have extended AuNWs to flexible electronics. Zhu et al. [32] found that AuNWs, which were created by integrating stretchable electrodes that developed vertically from patterned gold seeds in PDMS, shaped and formed an inherently stretchable organic polythiophene transistor series. Transistors using a dielectric gel electrolyte had excellent mechanical strength, even when a 100% strain was added, without noticeable decrease in efficiency [33].

Based on CNTs' outstanding mechanical features, many studies have sought to use them as stretchable electrodes. Various manufacturing processes, particularly spray coating and floating catalyst chemical vapor deposition, have been developed to produce such stretchable electrodes. Creating a polymer matrix by dispersing CNTs in a matrix material has also been tested. In this case, it is imperative to distribute the CNTs evenly [34]. The ionic liquids of single-walled carbon nanotubes have been spread in a fluorinated rubber control cup with the corresponding conductor, an active matrix stretchable organic light-emitting diode [4]. Wang et al. recently developed a transistor with an inherently stretchable array using CNTs as electrodes and an azide cross-linked SEBS dielectric (SEBS-X-azide). They manufactured a 10×10 matrix of touch sensors using CNT-based stretchable electrodes. The contact sensors in matrix form can be used as electronic skin [35].

1.2 WEARABLE ELECTRONICS

Due to the demand for virtual reality (VR) and augmented reality (AR), wearable technologies that can easily measure human activity have been extensively explored. Precise analysis of human motion is vital for successful interaction with a virtual environment. Finger measurements have been a central field of study for creating interactive experiences in VR and AR applications and have played a critical role in techniques for manipulating objects and communicating with the external world. Significant research has been carried out to test finger activity [36, 37]. The devices that have been developed can be graded as non-glove or glove—usually, multiple monitors containing reflective markers or X-ray imaging record human activity in non-glove structures. Mobility, however, is hampered by the fixed cameras and the need for peripherals [38]. Glove-based measuring devices, on the other hand, are more mobile. Glove-based systems are equipped with magnetic sensors, organic light-emitting diodes, potentiometers, and flex sensors. Although these devices will precisely determine finger joint angles, the normal finger movement is confined to the sensors' glove. While some devices can calculate 3D finger motions, they can have limitations in device complexity and precision of calculations. An extra module to measure the sensor position is required for the 3D magnetic sensor system, as shown to the right of Figure 1.1 [37, 39]. As an alternative approach for finger motion measurement, soft sensors have been used. Soft sensors are usually made from soft materials like silicone with embedded liquid-filled microchannels like EGaIn. The microchannels deform, causing a change in the leading liquid's electricity resistance if an external force is applied to the sensor. The applied intensity or pressure of the sensor may be determined by calculating the difference in resistance. The total device size can be minimized because soft sensors have an integrated structure. The elasticity of soft materials also decreases impairment of normal finger mobility.

1.3 WEARABLE ACTUATORS

There is a broad range of uses of wearable actuators, including for implantable, smart wearable watches for biomedical recovery and assistance in day-to-day life, bioinspired and biomimetic devices, and grasping and manipulating delicate objects to adaptable locomotives; wearable actuators are built of soft actuators constructed from active elastomeric materials [40]. This section presents a newly designed soft actuator and design tool to improve production methods for mechanically enhanced actuators for the above-mentioned applications. In recent years, thanks to the many advantages provided by the majority of soft materials—such as being lightweight and having low processing costs, large degrees of flexibility, and high adaptability—soft robotics has emerged as an area of study, integrating knowledge from various engineering disciplines, particularly materials science, chemistry, and robotics technology [41]. The use of flexible, autonomous robots and architectures in different areas—including bioinspired and biomimetic frameworks and adaptive locomotion in the unstructured landscape, motion detection and autonomous navigation, the seizure and handling of delicate materials, operative surgical instruments, and biomedical reconstruction—may be feasible [42–44].

Soft actuators constitute a vital aspect of soft robotic systems, and they can be operated in a variety of ways, such as through electric load actuation, chemical processes, storage alloys, and compressed liquids. Due to the ease of manufacture, safety of operation, high strength to weight ratio, and low cost, soft pneumatic actuators (SPAs) are especially desirable for wearable robotic systems. Usually, the actuators consist of air corridors and deformation chambers with atmospheric input pressure [41, 45]. SPAs may be used to produce various motion patterns like linear expansion, contracture, folding, and rotating motions, and also to introduce mechanical stresses or moments inside the desired scope, depending on the design and the arrangement of the actuator. The above renders them theoretically very useful in portable, human-assistive systems that can preserve or mimic body movement or aid everyday activities. There are many examples in the literature of beneficial, portable rehabilitation products powered by SPAs [46] (see Table 1.1).

1.4 WEARABLE SENSORS

Sensors may be categorized based on frequency output or the various methods they use for signaling pathways. Transduction can take place through a variety of methods. There are currently three main transduction approaches based on sensing mechanisms: (1) electrochemical, (2) optical, and (3) acoustic mechanical detection. In addition, steady progress is being made in the design and optimization of modern

TABLE 1.1
Some applications of flexible actuator nanoelectronics with the displacement angles controlling movement

Application	Actuator motion profile	Displacement requirements	References
Assistive glove	Bending	Bend up to 150° angle	[46, 47]
Trunk support belt	Linear tensile	40 mm at 200 kPa	[40, 48]
Neck support	Bending	Bend up to 90° angle	[49, 50]
Hip exosuit	Bending	Stretching up to 300 mm	[45, 51]
Ankle diagnostic device	Bending	Bend up to 90° angle	[52]

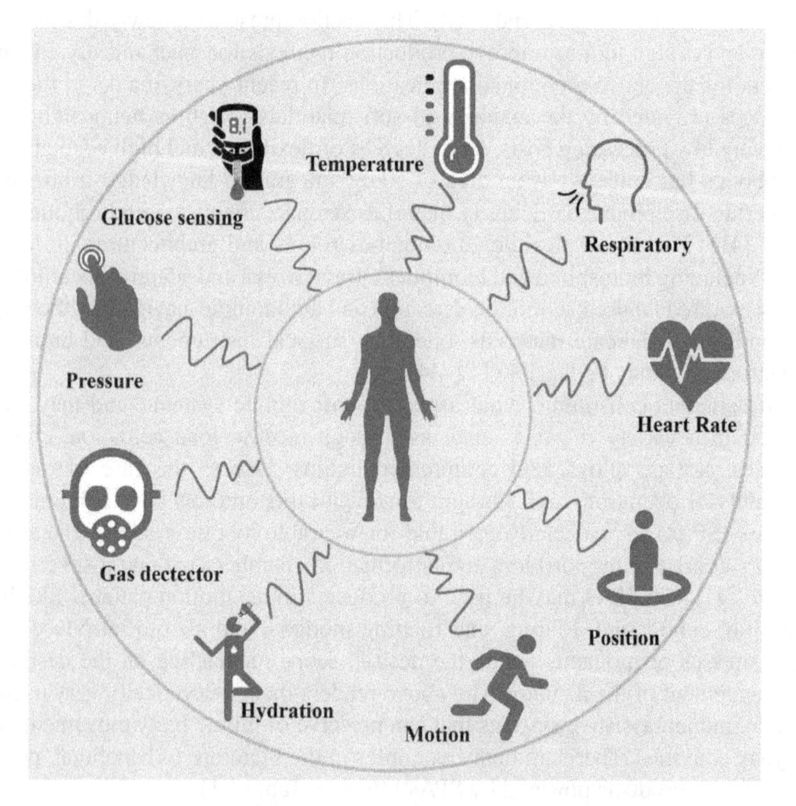

FIGURE 1.2 Applications of wearable nanosensors in health monitoring.

transducers' detection mechanisms to produce new sensor types. The three primary transduction methods have various subtypes, as presented in Figure 1.2. In conjunction with other methods, a variety of transduction systems are available. Several reports have been published on the architecture, classification, and optimization of extremely sensitive and complex nanosensors for indicators of interest. This area introduces a new age in disease control with, perhaps, more successful methods of treatment. The detection limit of nanosensors has been expanded to molar concentrations of pico (10^{-12}), femto (10^{-15}), and atto (10^{-18}).

1.5 ELECTRICAL SENSING

Many analytes can use electrochemical sensors. Electrical detection is a rapidly evolving area. Considering the electrochemical approach, there are currently some suggested marketable devices, such as for the detection of pathogens and poisons. The extraordinary characteristics of this approach include specificity, low cost, compatibility with current chip miniaturization and lab-on-a-chip techniques, minimal power requirements, and no pre-sampling (no impact of turbidity or sample color), and electrochemical detection is expected soon. A chemical reaction where electrons release, accept, or swallow ions is the standard mechanism of electrochemical nanosensors [53]. Electrochemical detection methods have various subtypes depending on the spectroscopic signal groups, such as potentiometry, voltammetry, and amperometry. Electrochemical analyses are often paired with immunoassays. The sensors are referred to as immune sensors. These approaches focus on the signal types production and amplification of signals.

Nanosensors based on electrical signal detection are the primary source of tunable properties and can be used for straightforward quantitative measurement, notably the recording of nano field-effect transistors (FETs). Rock-shaped nanomaterials (nanowires, nanorods, nanoribbons) are used for chemical nanosensors based on the FET concept. This results in a shift of impedance and creates a signal if targeted analysts stick to the active region. Nanostructures with these morphologies improve the susceptibility of and active region for current flow relative to the operation of flat detector surfaces [54].

1.6 OPTICAL SENSING

Nanosensor optical signals have high sensitivity due to the ability to distinguish between active sites of nanomaterials and light signals. However, exposure depends heavily on an optical phenomenon detection mode. Optical sensors are used for various forms of spectrometric analysis, such as absorption, visible radiation, fluorescence, Raman, solid-state relays, refraction, and equalitarian scattering [55].

The excitation of fluorescence with gold nanoparticles (AuNPs) is an example of optical signal translation. Fluorescence of isothiocyanates from fluorescein traveling through the AuNPs is positively exhausted, and there is no visible signal for radiation. The same molecule displays Raman scattering signals and hence functions as a Raman probe near these NPs. An empathetic identification of nanosensors depends on surface plasmon resonance (SPR), a traditional analysis-based technique that observes molecular forces. The refractive index fluctuates due to molecules joining onto small areas of the metal surfaces. The surface lighting event produces coherent electron surface oscillations susceptible to electromagnetic fluctuations on the boundary. The molecular binding events may cause these variations to lead to functionalities in the observed reflectivity spectrum. However, the achievement of high sensitivity is difficult since SPR usually lacks resolution as the non-specified binding interacts with the material efficiency.

Wearable and integrated medical devices have tremendous promise with garments and appliances with large dense surfaces and the prospect of inclusion via built-in sensors, pockets, and integrated electronics. Any use of traditional generic gel electrons places a variety of constraints on users; the user can feel uncomfortable, and they often result in skin inflammation as they allow unnecessary bacteria to develop. Also the electrons are not reusable and gradually degrade as the gel dries out. As a feasible option, dry electrodes might facilitate discrete positioning of electrodes on some regions of the wearer's body, allowing precise monitoring of physiological signals like heartbeat, the respiratory system (pulmonary and abdominal), electrodermal impedance, and galvanic skin responsiveness [56, 57]. Comprehensive and detailed knowledge about the wearer's well-being can be accessed using embedded textile drivers to connect various data acquisition nodes directly. Multiple sensors help computer processing to be minimized and allow for data fusion to provide more accurate information. Several technologies and techniques have been used to develop dry electrodes, such as knitted electrodes based on a smooth cutting and intarsia technology, woven electrodes produced by Jacquard technology, and polymer disposable electrodes with snap fastener connectors for attachment to the fabric [58–60].

Flexible touch sensors are grouped according to several sensing systems in the literature, including capacitive, piezoresistive, piezoelectric, thermoelectrical, triboelectric, and other sensing materials. A capacitive sensing system is typically preferred because of its basic structure, reliability, and temperature isolation. Cambridge University researchers introduced the standard capacitive sensor, comprising gold upper and bottom electrodes and a dielectric sheet of silicone rubber. The interval between gold electrodes decreased when an external force was exerted, and the capacitance changed accordingly. The device's pressure response was 1.4 %/N—reasonably low for practical use [61].

The use of lighter dielectric material is one approach that has been applied to improve the sensitivity of the lightweight tactile sensors. For example, Lee et al. [48] presented a capacitive, dielectric tactile sensor using air. Under a small force, it was much easier to deform and thus extremely susceptible. For a slight deflection, a sensitivity of 3%/mN was assessed. In this way, the operational range was limited because there was little force in the air distance. The integration of microstructures on the surface of dielectric layers is also a way of increasing adaptation. Liang et al. [62] introduced a versatile capacitive contact sensor integrated with a pyramid of PDMS as a type of dielectric layer that dramatically improves sensitivity to 67.2%/mN. It is expected that a new configuration and morphology of the microstructure will lead the tactile sensor to a different sensitivity, but few papers are available for thorough study. This chapter has contrasted four types of microstructures in the dielectric layer to enhance capacitive photocatalytic activity.

Different contact sensors have shown exciting tension sensing ability for potential implementation in robot skin [63]. Maiolino et al. specified a tactile sensor that is appropriate for robots and captures an intensity range of 0.4–10 N and spatial resolution of 1–2 mm [63]. Dahiya and Dario documented robot touch interface parameters that integrated human senses with sensitivity powers of 0.01 N to 10 N and progressive power resolutions of 0.01 N, 1 mm (fingers), and 5 mm (palm).

A signal processing circuit is needed in certain situations when the interface between the touch sensor and the robot power is used [63, 64]. For example, Chen and his research group transferred a versatile contact sensor to a prosthetic palm [36].

Wearable devices had already increasingly gained considerable interest due to their various applications for increased quality of life and ability to measure both body function and atmospheric conditions. Different wearable sensors have been registered. These sensing devices were quickly developed as a medium for non-conforming materials, such as the human body. Park et al. [65], among others, developed adaptable gas sensors with graphene/AgNW integrated functions (primary cause, discharge, interconnector, and antenna sprockets) that are stretchable. The graphene/AgNW hybrid gas sensors exhibit excellent sensor performance (sensitivity, reaction time, and restoration) and retain their efficiency in the stretched state (compressive stiffness: 20%).

Furthermore, Kim et al. [20] documented flexible contact lenses that use graphene/AgNW conductors to determine the existence of glucose and measure intraocular pressure. These have the advantage of high mechanical efficiency: the materials of the modules can be supported mostly on the soft contact lens and work in a stable manner with the electronics. Several wearable sensors have become available in the past few years that can measure different biosignals, and multi-detectable sensors have also been developed which can simultaneously measure several biosignals in one unit.

REFERENCES

1. Casson, A. J. 2015. *Ultra low power signal processing in mHealth: Opportunities and challenges.* 37th Annual International Conference of the IEEE Engineering in Medicine and Biology Society, Milan, Italy.
2. Amor, J. D., James, C. J. 2015. *Setting the scene: Mobile and wearable technology for managing healthcare and well-being.* In *2015 37th Annual International Conference of the IEEE Engineering in Medicine and Biology Society (EMBC)*, IEEE: 7752–7755.
3. Mannsfeld, S. C., Tee, B. C., Stoltenberg, R. M., Chen, C. V. H-H., Barman, S., Muir, B. V., Sokolov, A. N., Reese, C., Bao, Z. 2010. Highly sensitive flexible pressure sensors with microstructured rubber dielectric layers. *Nature Materials 9* (10), 859–864.
4. Sekitani, T., Yokota, T., Zschieschang, U., Klauk, H., Bauer, S., Takeuchi, K., Takamiya, M., Sakurai, T., Someya, T. 2009. Organic nonvolatile memory transistors for flexible sensor arrays. *Science 326* (5959), 1516–1519.
5. Jung, M., Kim, J., Noh, J., Lim, N., Lim, C., Lee, G., Kim, J., Kang, H., Jung, K., Leonard, A. D. et al. 2010. All-printed and roll-to-roll-printable 13.56-MHz-operated 1-bit RF tag on plastic foils. IEEE Transactions on Electron Devices, *57* (3), 571–580.
6. Steinhubl, S. R., Muse, E. D., Topol, E. J. 2015. The emerging field of mobile health. *Science Translational Medicine 7* (283), 283rv3.
7. Segev-Bar, M., Haick, H. J. 2013. Flexible sensors based on nanoparticles. *ACS Nano 7* (10), 8366–8378.
8. Xiao, F., Song, J., Gao, H., Zan, X., Xu, R., Duan, H. 2012. Coating graphene paper with 2D-assembly of electrocatalytic nanoparticles: A modular approach toward high-performance flexible electrodes. *ACS Nano 6* (1), 100–110.
9. Masala, O., Seshadri, R. 2004. Synthesis routes for large volumes of nanoparticles. *Annual Review of Materials Research 34*, 41–81.

10. Cushing, B. L., Kolesnichenko, V. L., O'Connor, C. 2004. Recent advances in the liquid-phase syntheses of inorganic nanoparticles. *Chemical Reviews 104* (9), 3893–3946.
11. Bhattacharya, S., Srivastava, A. 2003. Synthesis and characterization of novel cationic lipid and cholesterol-coated gold nanoparticles and their interactions with dipalmitoylphosphatidylcholine membranes. *Langmuir 19* (10), 4439–4447.
12. Haick, H. 2007. Chemical sensors based on molecularly modified metallic nanoparticles. *Journal of Physics D: Applied Physics 40* (23), 7173.
13. Daniel, M.-C., Astruc, D. 2004. Gold nanoparticles: Assembly, supramolecular chemistry, quantum-size-related properties, and applications toward biology, catalysis, and nanotechnology. *Chemical Reviews 104* (1), 293–346.
14. Zamborini, F. P., Bao, L., Dasari, R. 2012. Nanoparticles in measurement science. *Analytical Chemistry 84* (2), 541–576.
15. Qian, X., Park, H. S. 2010. The influence of mechanical strain on the optical properties of spherical gold nanoparticles. *Journal of the Mechanics and Physics of Solids 58* (3), 330–345.
16. Bae, S., Kim, H., Lee, Y., Xu, X., Park, J.-S., Zheng, Y., Balakrishnan, J., Lei, T., Kim, H. R., Song, Y. I. et al. 2010. Roll-to-roll production of 30-inch graphene films for transparent electrodes. *Nature Nanotechnology 5* (8), 574–578.
17. Yokota, T., Zalar, P., Kaltenbrunner, M., Jinno, H., Matsuhisa, N., Kitanosako, H., Tachibana, Y., Yukita, W., Koizumi, M., Someya, T. 2016. Ultraflexible organic photonic skin. *Science Advances 2* (4), e1501856.
18. Kim, J., Salvatore, G. A., Araki, H., Chiarelli, A. M., Xie, Z., Banks, A., Sheng, X., Liu, Y., Lee, J. W., Jang, K.-I. et al. 2016. Battery-free, stretchable optoelectronic systems for wireless optical characterization of the skin. *Science Advances 2* (8), e1600418.
19. Baca, A. J., Ahn, J. H., Sun, Y., Meitl, M. A., Menard, E., Kim, H. S., Choi, W. M., Kim, D. H., Huang, Y., Rogers, J. A. 2008. Semiconductor wires and ribbons for high-performance flexible electronics. *Angewandte Chemie International Edition 47* (30), 5524–5542.
20. Kim, J., Kim, M., Lee, M.-S., Kim, K., Ji, S., Kim, Y.-T., Park, J., Na, K., Bae, K.-H., Kim, H. K. 2017. Wearable smart sensor systems integrated on soft contact lenses for wireless ocular diagnostics. *Nature Communications 8* (1), 1–8.
21. Gong, S., Cheng, W. 2017. One-dimensional nanomaterials for soft electronics. *Advanced Electronic Materials 3* (3), 1600314.
22. Lagrange, M., Langley, D., Giusti, G., Jiménez, C., Bréchet, Y., Bellet, D. 2015. Optimization of silver nanowire-based transparent electrodes: Effects of density, size and thermal annealing. *Nanoscale 7* (41), 17410–17423.
23. Sun, H., Deng, J., Qiu, L., Fang, X., Peng, H. 2015. Recent progress in solar cells based on one-dimensional nanomaterials. *Energy & Environmental Science 8* (4), 1139–1159.
24. Rogers, J. A., Someya, T., Huang, Y. 2010. Materials and mechanics for stretchable electronics. *Science 327* (5973), 1603–1607.
25. Kim, K., Hyun, B. G., Jang, J., Cho, E., Park, Y.-G., Park, J.-U. 2016. Nanomaterial-based stretchable and transparent electrodes. *Journal of Information Display 17* (4), 131–141.
26. Zhang, Q., Liang, J., Huang, Y., Chen, H., Ma, R. 2019. Intrinsically stretchable conductors and interconnects for electronic applications. *Materials Chemistry Frontiers 3* (6), 1032–1051.
27. Amjadi, M., Pichitpajongkit, A., Lee, S., Ryu, S., Park, I. 2014. Highly stretchable and sensitive strain sensor based on silver nanowire–elastomer nanocomposite. *ACS Nano 8* (5), 5154–5163.

28. Tokuno, T., Nogi, M., Karakawa, M., Jiu, J., Nge, T. T., Aso, Y., Suganuma, K. 2011. Fabrication of silver nanowire transparent electrodes at room temperature. *Nano Research 4* (12), 1215–1222.
29. Garnett, E. C., Cai, W., Cha, J. J., Mahmood, F., Connor, S. T., Christoforo, M. G., Cui, Y., McGehee, M. D., Brongersma, M. L. 2012. Self-limited plasmonic welding of silver nanowire junctions. *Nature Materials 11* (3), 241–249.
30. Lee, J.-Y., Connor, S. T., Cui, Y., Peumans, P. 2008. Solution-processed metal nanowire mesh transparent electrodes. *Nano Letters 8* (2), 689–692.
31. Langley, D., Lagrange, M., Giusti, G., Jiménez, C., Bréchet, Y., Nguyen, N. D., Bellet, D. 2014. Metallic nanowire networks: Effects of thermal annealing on electrical resistance. *Nanoscale 6* (22), 13535–13543.
32. Zhu, B., Gong, S., Cheng, W. 2019. Softening gold for elastronics. *Chemical Society Reviews 48* (6), 1668–1711.
33. Zhao, Y., Zhai, Q., Dong, D., An, T., Gong, S., Shi, Q., Cheng, W. 2019. Highly stretchable and strain-insensitive fiber-based wearable electrochemical biosensor to monitor glucose in the sweat. *Analytical Chemistry 91* (10), 6569–6576.
34. Ma, W., Song, L., Yang, R., Zhang, T., Zhao, Y., Sun, L., Ren, Y., Liu, D., Liu, L., Shen, J. et al. 2007. Directly synthesized strong, highly conducting, transparent single-walled carbon nanotube films. *Nano Letters 7* (8), 2307–2311.
35. Wang, S., Xu, J., Wang, W., Wang, G.-J. N., Rastak, R., Molina-Lopez, F., Chung, J. W., Niu, S., Feig, V. R., Lopez, J. et al. 2018. Skin electronics from scalable fabrication of an intrinsically stretchable transistor array. *Nature 555* (7694), 83–88.
36. Wang, Y., Liang, G., Mei, D., Zhu, L., Chen, Z. 2016. A flexible capacitive tactile sensor array with high scanning speed for distributed contact force measurements. In *2016 IEEE 29th International Conference on Micro Electro Mechanical Systems (MEMS), Shanghai, China*, 854–857.
37. Park, Y., Lee, J., Bae, J. 2014. Development of a wearable sensing glove for measuring the motion of fingers using linear potentiometers and flexible wires. *IEEE Transactions on Industrial Informatics 11* (1), 198–206.
38. Kuo, L.-C., Su, F.-C., Chiu, H.-Y., Yu, C.-Y. 2002. Feasibility of using a video-based motion analysis system for measuring thumb kinematics. *Journal of Biomechanics 35* (11), 1499–1506.
39. Kessler, G. D., Hodges, L. F., Walker, N. 1995. Evaluation of the CyberGlove as a whole-hand input device. ACM Transactions on Computer-Human Interaction *2* (4), 263–283.
40. Ilievski, F., Mazzeo, A. D., Shepherd, R. F., Chen, X., Whitesides, G. M. 2011. Soft robotics for chemists. *Angewandte Chemie International Edition 123* (8), 1930–1935.
41. Tolley, M. T., Shepherd, R. F., Mosadegh, B., Galloway, K. C., Wehner, M., Karpelson, M., Wood, R. J., Whitesides, G. M., 2014. A resilient, untethered soft robot. *Soft Robotics 1* (3), 213–223.
42. Marchese, A. D., Katzschmann, R. K., Rus, D. 2015. A recipe for soft fluidic elastomer robots. *Soft Robotics 2* (1), 7–25.
43. Laschi, C., Cianchetti, M., Mazzolai, B., Margheri, L., Follador, M., Dario, P. 2012. Soft robot arm inspired by the octopus. *Advanced Robotics 26* (7), 709–727.
44. Kim, S., Laschi, C., Trimmer, B. 2013. Soft robotics: A bioinspired evolution in robotics. *Trends in Biotechnology 31* (5), 287–294.
45. Polygerinos, P., Wang, Z., Overvelde, J. T., Galloway, K. C., Wood, R. J., Bertoldi, K., Walsh, C. J. 2015. Modeling of soft fiber-reinforced bending actuators. *IEEE Transactions on Robotics 31* (3), 778–789.

46. Agarwal, G., Besuchet, N., Audergon, B., Paik, J. 2016. Stretchable materials for robust soft actuators towards assistive wearable devices. *Scientific Reports 6* (1), 34224.
47. Choi, S., Han, S. I., Kim, D., Hyeon, T., Kim, D. H. 2019. High-performance stretchable conductive nanocomposites: Materials, processes, and device applications. *Chemical Society Reviews 48* (6), 1566–1595.
48. Li, T., Li, Y., Zhang, T. 2019. Materials, structures, and functions for flexible and stretchable biomimetic sensors. *Accounts of Chemical Research 52* (2), 288–296.
49. Liu, Y., Liu, Z., Zhu, B., Yu, J., He, K., Leow, W. R., Wang, M., Chandran, B. K., Qi, D., Wang, H. et al. 2017. Stretchable motion memory devices based on mechanical hybrid materials. *Advanced Materials 29* (34) 1701780.
50. Liu, Y., Wang, H., Zhao, W., Zhang, M., Qin, H., Xie, Y. 2018. Flexible, stretchable sensors for wearable health monitoring: Sensing mechanisms, materials, fabrication strategies and features. *Sensors (Basel) 18* (2), 645.
51. Matsuhisa, N., Chen, X., Bao, Z., Someya, T. 2019. Materials and structural designs of stretchable conductors. *Chemical Society Reviews 48* (11), 2946–2966.
52. Rogers, J. A., Someya, T., Huang, Y. 2010. Materials and mechanics for stretchable electronics. *Science 327* (5973), 1603–1607.
53. Zhu, S., Song, Y., Zhao, X., Shao, J., Zhang, J., Yang, B. 2015. The photoluminescence mechanism in carbon dots (graphene quantum dots, carbon nanodots, and polymer dots): Current state and future perspective. *Nano Research 8* (2), 355–381.
54. Kuila, T., Bose, S., Khanra, P., Mishra, A. K., Kim, N. H., Lee, J. H. 2011. Recent advances in graphene-based biosensors. *Biosensors and Bioelectronics 26* (12), 4637–4648.
55. Stern, E., Vacic, A., Reed, M. A. 2008. Semiconducting nanowire field-effect transistor biomolecular sensors. *I*EEE Transactions on Electron Devices *55* (11), 3119–3130.
56. Finni, T., Hu, M., Kettunen, P., Vilavuo, T., Cheng, S. 2007. Measurement of EMG activity with textile electrodes embedded into clothing. *Physiological Measurement 28* (11), 1405–1419.
57. Villarejo, M. V., Zapirain, B. G., Zorrilla, A. M. 2012. A stress sensor based on galvanic skin response (GSR) controlled by ZigBee. *Sensors (Basel) 12* (5), 6075–6101.
58. Trindade, I. G., Martins, F., Baptista, P. 2015. High electrical conductance poly (3, 4-ethylenedioxythiophene) coatings on textile for electrocardiogram monitoring. *Synthetic Metals 210*, 179–185.
59. Chen, Y.-H., De Beeck, M. O., Vanderheyden, L., Carrette, E., Mihajlović, V., Vanstreels, K., Grundlehner, B., Gadeyne, S., Boon, P., Van Hoof, C. 2014. Soft, comfortable polymer dry electrodes for high quality ECG and EEG recording. *Sensors (Basel) 14* (12), 23758–23780.
60. Tada, Y., Amano, Y., Sato, T., Saito, S., Inoue, M. 2015. A smart shirt made with conductive ink and conductive foam for the measurement of electrocardiogram signals with unipolar precordial leads. *Fibers 3* (4), 463–477.
61. Muhammad, H., Oddo, C., Beccai, L., Recchiuto, C., Anthony, C., Adams, M., Carrozza, M., Hukins, D., Ward, M. C. L. 2011. Development of a bioinspired MEMS based capacitive tactile sensor for a robotic finger. *Sensors and Actuators A: Physical 165* (2), 221–229.
62. Liang, G., Wang, Y., Mei, D., Xi, K., Chen, Z. 2015. Flexible capacitive tactile sensor array with truncated pyramids as dielectric layer for three-axis force measurement. Journal of Microelectromechanical Systems *24* (5), 1510–1519.
63. Maiolino, P., Maggiali, M., Cannata, G., Metta, G., Natale, L. 2013. A flexible and robust large scale capacitive tactile system for robots. *IEEE Sensors Journal 13* (10), 3910–3917.

64. Dahiya, R. S., Valle, M., Metta, G., Lorenzelli, L. 2007. POSFET based tactile sensor arrays. In *2007 14th IEEE International Conference on Electronics, Circuits and Systems*, IEEE: 1075–1078.
65. Park, J., Kim, J., Kim, K., Kim, S.-Y., Hyung Cheong, W., Park, K., Song, J. H., Namgoong, G., Kim, J. J., Heo, J. et al. 2016. Wearable, wireless gas sensors using highly stretchable and transparent structures of nanowires and graphene. *Nanoscale*, *8* (20), 10591–10597.

2 Graphene-Based Materials and Devices for Transparent Stretchable Electronics

Jyothy G. Vijayan

CONTENTS

2.1 INTRODUCTION

Flexible and stretchable electronics have emerged as a novel material for a variety of applications due to their benefits for daily life. Strong efforts have been made to fabricate stretchable electronics due to the high demand for intelligent, wearable, and integrated electronic systems. Mechanical compliance is crucial in the making of transparent stretchable electronic devices. Stretchable devices should not incur physical strain or damage under stretching conditions [1, 3]. Graphene, a 2D material, is highly applicable as a conventional semiconductor material and also in the synthesis of flexible electronics [7, 9]. It is considered the best among stretchable materials due to its high tensile strength and fracture at low strain. Graphene-based flexible devices have advantages over rigid ones due to their improved comfort, greater space efficiency, and better durability [22]. For rigid devices, the fabrication process is limited to conventional methods. But in the case of flexible and elastomeric 2D materials like graphene, boron nitride and transition metal dichalocogenides have much potential due to their electrical, chemical, optical, and mechanical properties [12], which make them ideal for flexible and stretchable electronics.

Graphene possesses strong optoelectronic and mechanical properties and also has good adhesion characteristics. Graphene is used in organic light-emitting diodes and organic photovoltaics. Graphene exhibits higher specific volume ratio due to its higher chemical stability. It is also used in the doping process and in high-performance power batteries. Graphene has outstanding electrical, chemical, mechanical, and optical properties for macroscopic applications, which makes it suitable for use as a transparent conducting film [21]. Due to these characteristics, graphene has replaced indium tin oxide (ITO) films. Macroscopic applications include energy storage, displays, solar cells and transparent heaters. Large-scale graphene films have been synthesized using chemical vapor deposition (CVD). Because of its exceptional optoelectronic characteristics and conductivity, graphene is in high demand as a stretchable material in electric heater preparation. Due to its high performance, graphene has also been used as a high-performance supercapacitor in electronic devices and portable sustainable self-charging power systems. This review details the recent synthesis and application of graphene-based stretchable and transparent devices.

2.2 STRETCHABLE ELECTRONICS: METHODS AND MECHANISMS

Mechanical buckling using the “wavy structure configuration method”, mesh structures, and island interconnect configurations are used to configure flexible and stretchable structures [14]. In the island interconnect method, stretchable interconnects are prepared using highly malleable electronic materials (e.g. liquid metals) by out-of-plane of deformation pop-up by interconnectors. Now 2D geometries are used in the fabrication of stretchable materials and in spiral interconnect arrays. Using the interconnect island method and mechanism, an effective mesh structure produces a large surface that is reversible in nature and offers high stretchability for electronic circuits. Fractal-inspired structural training is used to create the conductive interconnect configurations. Fractal configurations are highly efficient in transforming inextensible sheet to super-conformable and high-tensile sheet. Usually rigid, thin films are combined with a stretchable substrate [14]. Achieving wide application as a stretchable device is a challenge, however.

Origami, an old paper folding art, has been applied in the making of stretchable electronic devices [13]. It is in great demand for future portable display and

TABLE 2.1
Different methods and their properties in graphene-based devices

Methods	Uses
Drop casting	Placing a droplet on the substrate Formation of thick film Casting is not repeatable
Spin coating	Repeatable casting Produces uniform fill
Doctor blading	Constant movement between blade and substrate Formation of gel layers
Dip coating	Good uniformity and drying Thin layer and area coverage
Spray coating	Formation of thin film Electrohydrodynamic atomization

wearable electronics. It fits the need of foldable electronics, and its fractal designs are attractive for future-generation portable displays and wearable electronics. Origami designs are used for the synthesis of foldable conductors, stretchable electronics, and supercapacitors. Recent trends in miniaturized wearable electronics require highly flexible, stretchable, and portable energy storage devices. Kirigami is a version of origami which combines strategic folding and cutting. The main advantage of kirigami is that it can transform extensible substrates into highly tensile ones. Kirigami designs can improve the stretchability of materials. Kirigami sheets can exhibit out-of-plane mechanical deformation, and the sheet is highly stretchable. Kirigami structures are also used in the manufacture of stretchable triboelectric nanogenerators (NGs), which are used in power generators and self-powered senior applications. Kirigami is useful for micro and nanoscale applications such as bioelectronics and energy storage devices, bendable and foldable optoelectronic devices and microrobots, etc.

Direct printing methods, especially solution coating and deposition, are the fabrication techniques used for kirigami designs. Printing is considered useful due to the low cost of electronic devices, large-scale production, etc. [11]. Contact printing and non-contact printing (i.e. aerosol jet printing) are the printing methods used in making stretchable devices. The main disadvantage of the non-contact printing method is that nozzles can become clogged. Solution coating and deposition methods include spin coating, drop casting, doctor blading, dip coating, and spray coating (Table 2.1).

2.3 CONVENTIONAL STRETCHABLE ELECTRONICS

Electrodes and conductors are materials that permit the flow of current in different directions. Soft and stretchable conductive materials are made using intrinsically flexible materials as conductive reinforcers [15].

Stretchable electronic conductors like Ag nanoflakes, Au-Ni nanoparticles, cotton nanowires, and metal nanoparticles are used as potential fillers in an elastomer matrix to form stretchable structural designs [8]. Here it is observed that the stretchability in the devices—achieved by different techniques like solution processing methods, low temperature, sintering, etc.—helps to ensure conductivity. Photolithography also helps in the fabrication of silver nanowire (AgNW) networks in polydimethylsiloxane (PDMS) like polymeric materials. Stretchable conductors are very useful in wearable and stretchable electronics, energy storage, and artificial skin synthesis.

In stretchable printing tracks, printing techniques are very cost effective for the recent trend of electronic devices and stretchable circuits. The designing, structuring, and patterning of electrically transparent and conductive materials on stretchable substrates results in devices with good mechanical characteristics—being bendable, stretchable, foldable, etc. Printing ink should adhere to the substrate without any pre-treatment. To enable this, some techniques have been identified to build a good bond between conductive inks and the surface of the elastomers. Inkjet printing, roll-to-roll printing, and screen printing have been optimized for this purpose. Designing and printing on stretchable substrates simultaneously and directly remains a big problem to be solved. Recently, researchers have found that fabrication of Ag-doped PDMS by screen printing enhanced stretching to 20% [19]. Recently, writable free inks have been synthesized for stretchable electronics with high conductivity. Stretchable heaters are an extension of stretchable conductors. In flexible heaters, the temperature is regulated and controlled by raising the temperature of electrodes and resistors. These heaters are used in thermal management and healthcare aids. Uniform healing of stretchable heaters has been achieved by adding dopants so that nanowires are uniformly or homogeneously dispersed in the elastomeric media.

Two strategies have been adopted for stretchable energy conversion and storage: (1) applying the stretchable functional materials directly and (2) utilizing the above-mentioned structural designs under stress and mechanical strain. Advances in stretchable energy conversion and storage devices include solar cells, supercapacitors, battery NGS, and triboelectric NGs. Stretchable solar cells include thin film solar cells and dye-sensitized solar cells. The advantage of using solar cells is that organic silicon solar cells can reach 64.4% compactness. The limitations of using organic solar cells is their low stretching properties, high cost, poor performance of organic photovoltaics, and lower environmental stability. Inorganic solar cells are efficient for stretchable solar cells because of their mechanical properties. Elastic solar cells are more efficient due to the resistance to fracture under strain. It is very difficult to synthesize supercapacitors with high stretchability, durability, compressibility, and twistability. In supercapacitor synthesis, solid state electrolytes and separators are assembled in two electrodes. The goal is to achieve good stability and high specific stretchability. There is a new trend to make omnidirectional stretchable supercapacitors with high flexibility. Due to high cost, a complex manufacturing process, and low stretchability, the existing use of stretchable supercapacitors is limited. Stretchable batteries are fabricated using kirigami designs. Zinc-based batteries are more efficient due to their air stability, inexpensive methods, and combination of transparency, flexibility, and stretchability. They have attracted lots of interest due to the practical value of wearable electronics in everyday life. Recently, transparent stretchable electronics have met the requirements of patchable, wearable, or implantable electronics. Transparent stretchable electronics have promising application in disease diagnostics and health monitoring [10].

2.4 GRAPHENE-BASED STRETCHABLE MATERIALS

New materials provide an alternative future for stretchable electronics. Conductive rubber-based elastomers filled with carbon black resist strain are the conventional materials used in stretchable electronics. It has been found that any material in thin form is flexible because of its bending strain, which decreases linearly with thickness [21]. Nanoscale silicon in ribbons, wires, or membranes is flexible, but silicon wafer is brittle in nature. In 2004, the first single-layer graphene was isolated from graphite. 3D graphite contains different layers of 2D graphene held together by Van der waals forces. Graphene exhibits high surface to volume ratio, which helps sustain chemical stability. It also helps when adding graphene in the making of high-performance power storage devices. The mechanical exfoliation process gives monolayered graphene with a high crystalline structure with higher electrical characteristics. The solution-based method also helps to synthesize single-layer graphene through chemical exfoliation. The process requires strong acids and oxidants to gain a sheet of graphene oxide from a graphite-powder-dispersed solution. Chemical exfoliation is a solution-based method which can produce large quantities of graphene at low cost. This process helps to produce good-quality graphene with a high surface area.

Another effective method of graphene synthesis is thermal CVD. In this method, decomposites of hot hydrocarbons sources on catalytic surfaces were observed. Graphene possesses a unique band structure which enhances the properties of its electrons so that it performs 10–100 times better than other elements. Scientists have found that single-layer graphene synthesized by the mechanical exfoliation method displayed higher carrier mobility at room temperature. Graphene exhibits electrical resistance on stretching. ITO is used mainly as a transparent conductor for optoelectronic apparatus. But due to its low mechanical properties and cracking easily in strain, it has been replaced by other materials. These problems of ITO mean that an alternative to graphene has to be found which is an efficient transparent conductor for photonics and optoelectronic applications. The surface structure of graphene is similar to carbon nanotubes. Each possess a hydrophobic surface and can be doped by chemical treatment.

2.5 SYNTHESIS OF GRAPHENE

CVD mixed with transfer techniques and chemical exfoliation has been used in graphene-based electronic devices. To synthesize CVD graphene with high crystallinity, it is essential to plan, investigate, control, and enhance the size of graphene for higher applications. The CVD method is used to develop large surface area and high-quality graphene films on metals using gaseous carbon form. Ni, Pt, Co, Cu, etc. are used as catalysts for developing graphene-based electronics. To produce high-crystalline CVD graphene, it is essential to maximize the size of the graphene. Recently developed methods like plasma-enhanced CVD have advantages over traditional CVD methods. Plasma-enhanced CVD requires low processing temperatures, which helps graphene to grow on plastic substrate on nanostructures [24].

2.5.1 Graphene Flakes: Synthesis and Fabrication

Synthesis of graphene flakes from graphite in a solvent medium involves oxidation of graphite using strong acids and oxidants, chemical exfoliation of graphite layers via ultrasonication, and purification of the flakes through ultracentrifugation. The final product, graphene oxide (GO), is insulated and electrically conductive after the chemical treatments. Metal salts are highly effective intercalating agents for exfoliating graphite in low-cost processes, which helps to produce good-quality graphene flakes [23]. The synthesized flakes exhibit a large surface area and uniform deposition (Figure 2.1). Inkjet and spray painting techniques help to disperse the flakes in solvents.

2.6 APPLICATIONS OF GRAPHENE-BASED FLEXIBLE AND STRETCHABLE ELECTRONICS

Recently developed graphene composite materials for electronic devices are more flexible and stretchable than those that use conventional graphene and graphene derivatives (Figure 2.2). Graphene and graphene-based composites have proved to be outstanding materials for flexible and stretchable electronics (Table 2.2).

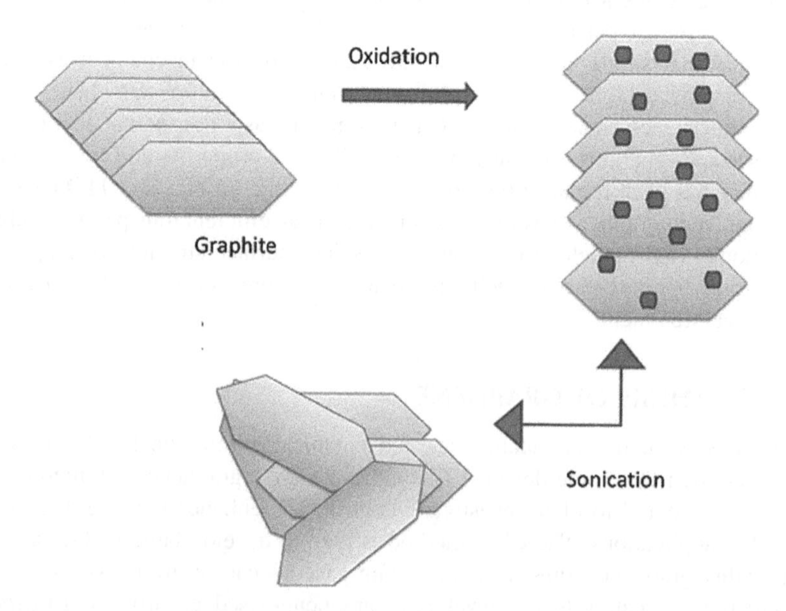

FIGURE 2.1 Graphene flakes generated from graphene.

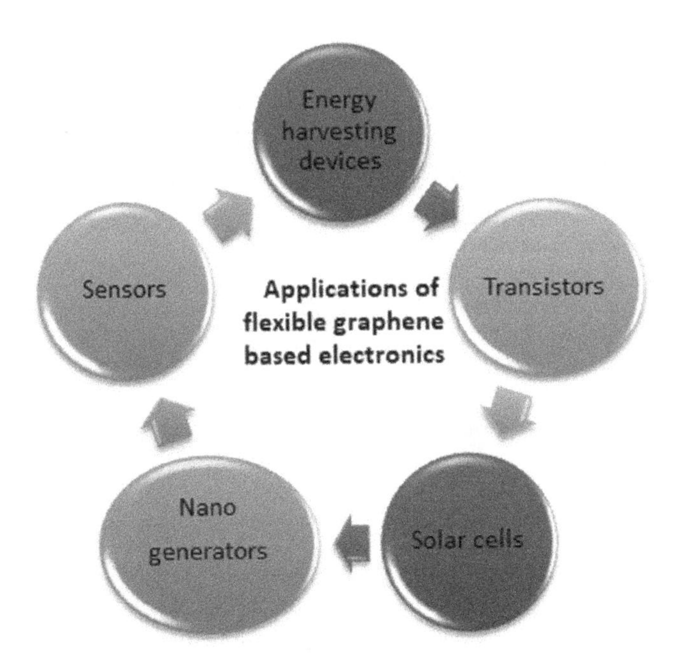

FIGURE 2.2 Applications of graphene-based flexible and stretchable electronics.

TABLE 2.2
Materials, methods, and stretchability of graphene and graphene-based composites

Materials	Methods	Stretchability	References
Graphene	Transferred film	6% up to the strain	[38a]
Graphene/PEDOT.PSS	Spin coating	15% up to the strain	[28]
Graphene/AgNW	Spin coating	100% negligible resistance	[42]
Graphene/AgNW	Spin coating	100% in tensile strain	[21]
AgNWs	Drop casting	Strong as high as 140%	[43]

2.6.1 In Electronics

2.6.1.1 Synthesis of Soft and Transparent Electrodes

Graphene is used as a flexible electrode to make devices such as light-emitting diodes (LEDs), energy convertors, energy storage devices, displays, sensors, etc. (Figure 2.3). Graphene also possess positive electrical properties such as low carrier concentration, high field effect mobility, good transmittance, and high mechanical flexibility. Graphene has replaced ITO, which has several disadvantages like high cost, low strain, and high corrosion. Graphene flakes are an efficient material for the preparation of flexible electrodes [5]. Reduced graphene oxides are used to make thin

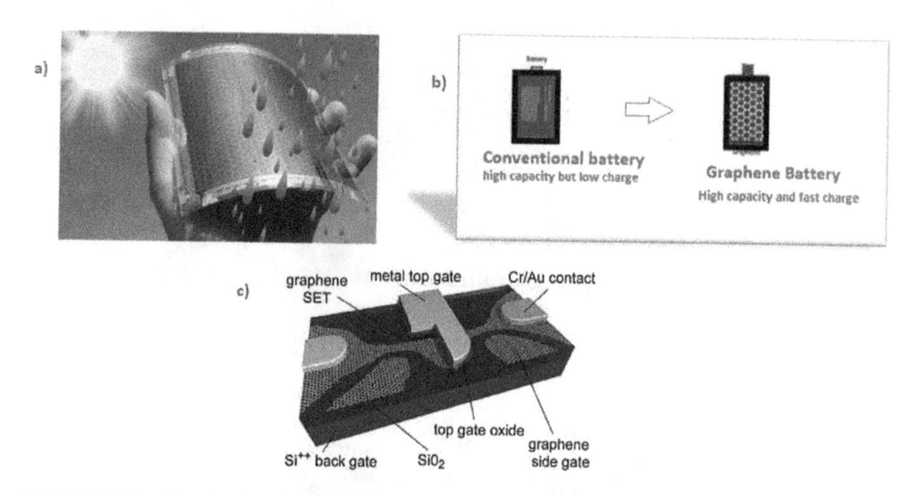

FIGURE 2.3 (a) Graphene-based solar cells, (b) graphene battery and (c) graphene-based generator.

film transistors on various substrates. But the main limitation of GO-based electrodes is incompatibility between high conductivity and transmittance.

2.6.1.2 Graphene-Based Flexible and Top-Gated Stretchable Transistors

Graphene transistors exhibit novel electrical properties and fast operating speed due to their massless Dirac fermions and Berry phase. Major studies have been conducted on the intrinsic properties of graphene on Si in a bottom-gated transistor configuration. Bottom-gated graphene transistors are effective for enhancing the electrical properties on the transistors at low temperature and high magnetic field. Tin oxide and aluminum oxide, like conventional gate dielectric insulators, cannot be used in graphene transistors because of their low mechanical strength. Recently, researchers reported stretchable graphene transistors which show high performance [38a, 38b]. The transistors can work in low voltage with excellent mechanical flexibility and environmental stability.

2.6.1.3 Structure of Graphene with Different 2D Materials

Graphene is highly recommended for soft electronic devices. The limitation of using graphene in soft electronic devices is its negligible or zero band gap. It takes huge effort to generate the required band gap in the material, and this results in degradation of the material's electrical properties. Combination of graphene with other 2D materials like boron nitride can produce devices that are stretchable, thin, and mechanically stable in nature. Those heterostructures have lots of applications as flexible materials.

2.6.1.4 Graphene-Based Flexible Logic Devices

Conventional stretchable logic electronics made from carbon nanomaterials (CNTs), polymers, or oxide materials are the main components that could really benefit from advances in stretchable devices [2, 33]. Fabrication using these materials still has critical challenges to overcome. Graphene possess good atomic bonding and mechanical stability as well as a good Young's modulus range. It also exhibits high electrical characteristics. These characteristics are accompanied by stretchability and conformal features. Those properties make graphene a promising flexible material in electronics. Graphene which is atomically very thin displays high stretchability due to hexagonal lattice arrangements in its structures with strong carbon bonds. Graphene logic gates have been fabricated by combining the two; that is, graphene transistors with homogeneous characteristics. Graphene is used here due to its optical transmittance nature and high flexibility.

2.6.2 Graphene-Based Energy Storage Devices

In the field of flexible and stretchable electronics, there is great demand for development of energy-harvesting and energy storage units. There is a need to make current devices, light, small, stretchable batteries, etc.

a) **Lithium ion battery:** Compared to other carbon material, graphene has the best electrode potential with good surface area and Young's modulus.
b) **Supercapacitors:** Flexible supercapacitors are attractive because of their high power density and good mechanical properties [36, 29]. Multilayer GO-based planar supercapacitors have a high surface area [4].
c) **Energy convertors:** Studies have been conducted to transform solar and mechanical energy to electrical signals in order to construct a renewable source of energy and power [32]. Silicon photovoltaic technology is mainly used in conventional devices. This has disadvantages like high cost and a solution-based process; that is, flexible organic photovoltaics doped with a double layer of graphene.

2.6.3 Nanogenerators

Nanogenerators are used in flexible and stretchable electronics. They are used in biosensors and body-implanted devices. CVD-grown graphene is usually used in transparent piezoelectric energy-harvesting NGs. 2D graphene-based lead zirconate titanate NGs display high mechanical stability, bendability, and durability, and perform well. NGs are made using nanoscale piezoelectric materials and thin film insulators with high piezoelectric constants. Graphene-based NGs display good optical and mechanical properties. They provide flexible and transparent device applications. Graphene-assisted NGs can be considered an important part of stretchable electronics. But to improve the commerciality of NGs, triboelectric graphene have to address weak adhesion between graphene electrodes and low resistance to strains [44 and 6]. Graphene exhibits good performance as a stretchable and flexible transparent electrode while connected with inorganic piezoelectric material [42]. But mechanical durability is considered the main limitation facing the devices. Recently, researchers developed polymer piezoelectric material with high flexibility, optical transparency, and large surface area [20 and 34]. P-doped graphene-based NGs are produced by large-scale electronic devices that roll the material up to 100 times.

2.6.4 Stretchable and Ultra-Transparent Graphene Electrodes

Graphene displays superior properties like high atomic thickness, high transparency, and conductivity than CNT [16, 18]. Due to its highly flexible nature, graphene has been used as flexible electronics in different devices. But a limitation of using graphene in stretchable devices is its tendency to crack at small strains [37]. Graphene is used in electrodes for omnidimensional heterostructural semiconductors like those used in 1D, 2D, and omnidirectional quantum dots, solar cells, and LEDs.

Transistors are very important in flexible and stretchable applications as they provide sensor readouts and signal analyses. We can prepare highly stretchable graphene by integrating graphene scrolls, and we can also make highly transparent transistors using multilayer graphene/graphene scrolls.

2.6.5 Solar Cells

Many researchers have developed flexible organic solar cells with reduced GO or CVD-based graphene. CVD-based graphene exhibits lower sheet resistance than reduced GO. Flexible organic solar cells with graphene oxide performs better than ITO-based organic solar cells. This is mainly due to the fit factor of the organic solar cells in ratio with the layers. Recently, many studies have been conducted using flexible solar cells with graphene electrodes [27]. The performance of ITO-based organic solar cells is comparable to graphene-based ones. But the disadvantage of using ITOs are (a) the poor film coating of PEDOT.PSS (poly(3,4-ethylenedioxythiophene) polystyrene sulfonate) and (b) the lack of chemical stability of dopants. Moisture and oxygen also react with dopants and decrease their chemical stability.

2.6.5.1 Graphene-Based Organic and Inorganic LEDs

Graphene is the material with most potential to make flexible LEDs, due to its use as an anode and charge transport layer. Graphene is used to make transparent conductive electrodes [15].

2.6.6 Sensors

3D stretchable nanopaper with nanocellulose and graphene can overcome the strain generated by human joints and can be used as strain sensors [25]. Nanopaper-based strain sensors are used to detect the movement of fingers at a frequency of 1 Hz. Tactile sensors which possess high sensitivity can be used as an effective sensor for e-skin sensing systems [28]. Recently, laser-scribed graphene tactile sensors have been used efficiently. Laser-scribed graphene sensors can determine a wide pressure range with high sensitivity. Touch sensors and chemical sensors have received attention in the health sector and in relation to flexible wearable electronics [30]. Optical,

TABLE 2.3
Sensitivity range of graphene-based stretchable devices

Devices	Response type	Sensitivity range (%)	References
Nanopaper	Linear	100	[38b]
Foam	Exponential	77	[14]
Ripple	Linear	30	[35a]
Ribbon	Linear	3	[17]
Glove sensor	Linear	2	[2]
Woven fabric	Exponential	30	[35b]

chemical, electrical, and effective mechanical properties make graphene an excellent touch sensor material [39]. Methods used in the making of graphene-based sensors include rosette gauge arrangement, foam, percolate film, and microstructure arrays. 3D tactile sensors are preferred because of their high sensitivity in low pressure parts. Pressure output is divided into two: low pressure (0–100 Pa) and a saturated region value (> 100 Pa). For the synthesis of e-skin sensors, high-sensitivity tactile sensors are preferred. Graphene-based touch sensors are more efficient than ITO due to their high mechanical stability and flexible nature. Sensors that can identify and track the movement of the human body are an emerging novel class of electronics [45, 26]. Graphene-based sensors can identify the extent and area of pressure applied. They have a defect-free transfer method and good uniformity, chemical stability, and long-term stability of dopants in air (Table 2.3).

2.7 CONCLUSIONS AND FUTURE ISSUES

Graphene exhibits positive mechanical, chemical, optical, and physical properties which are highly desirable in making flexible and stretchable materials. Graphene has good applicability in areas such as NGs, electrodes, transistors, touch panels, and actuators. To make graphene-based stretchable devices, one has to overcome critical challenges like chemical stability and durability of the materials. Graphene doped with other 2D materials with hybrid structures has more advantages for flexible and stretchable materials.

Graphene-based stretchable electronics have emerged because of their efficiency to coordinate and integrate with different functional materials and curvilinear surfaces [31, 46]. To develop stretchable power resources, existing rigid constituents should be replaced by stretchable structures for heterogeneous integration of hard and deformable material. Fabrication cost is a crucial factor in the preparation and application of graphene-based stretchable electronics [40, 41]. Chemical, optoelectronic, and physical properties of graphene make it suitable for the synthesis of flexible and stretchable devices. To overcome all its limitations, a new strategy for producing highly conductive graphene with good uniformity and chemical stability must be found.

REFERENCES

1. Ahn, J. H., & Je, J. H. 2012. Stretchable electronics: Materials, architectures and integrations. *Journal of Physics D: Applied Physics*, 45(10): 103001.
2. Bae, S.-H., Lee, Y., Sharma, B. K., Lee, H.-J., Kim, J.-H., & Ahn, J.-H. 2013. Graphene-based transparent strain sensor. *Carbon*, 51, 236–242.
3. Cai, S., Han, Z., Wang, F., Zheng, K., Cao, Y., Ma, Y., & Feng, X. 2018. Review on flexible photonics/electronics integrated devices and fabrication strategy. *Science China Information Sciences*, 61(6): 060410.
4. Chee, W. K., Lim, H. N., Zainal, Z., Huang, N. M., Harrison, I., & Andou, Y. 2016. Flexible graphene-based supercapacitors: A review. *The Journal of Physical Chemistry C*, 120(8), 4153–4172.
5. Chen, D., Liang, J., & Pei, Q. 2016. Flexible and stretchable electrodes for next generation polymer electronics: A review. *Science China Chemistry*, 59(6), 659–671.
6. Chen, H., Xu, Y., Bai, L., Jiang, Y., Zhang, J., Zhao, C., ... & Gan, Q. 2017. Crumpled graphene triboelectric nanogenerators: Smaller devices with higher output performance. *Advanced Materials Technologies*, 2(6): 1700044.
7. Das, T., Sharma, B. K., Katiyar, A. K., & Ahn, J. H. 2018. Graphene-based flexible and wearable electronics. *Journal of Semiconductors*, 39(1): 011007.
8. Han, F., Su, X., Huang, M., Li, J., Zhang, Y., Zhao, S., ... & Wong, C. P. 2018. Fabrication of a flexible and stretchable three-dimensional conductor based on Au–Ni@ graphene coated polyurethane sponge by electroless plating. *Journal of Materials Chemistry C*, 6(30), 8135–8143.
9. Han, T. H., Kim, H., Kwon, S. J., & Lee, T. W. 2017. Graphene-based flexible electronic devices. *Materials Science and Engineering: R: Reports*, 118, 1–43.
10. Hong, Y. J., Jeong, H., Cho, K. W., Lu, N., & Kim, D. H. 2019. Wearable and implantable devices for cardiovascular healthcare: From monitoring to therapy based on flexible and stretchable electronics. *Advanced Functional Materials*, 29(19): 1808247.
11. Huang, Q., & Zhu, Y. 2019. Printing conductive nanomaterials for flexible and stretchable electronics: A review of materials, processes, and applications. *Advanced Materials Technologies*, 4(5): 1800546.
12. Jang, H., Park, Y. J., Chen, X., Das, T., Kim, M. S., & Ahn, J. H. 2016. Graphene-based flexible and stretchable electronics. *Advanced Materials*, 28(22), 4184–4202.
13. Ji, S., Hyun, B. G., Kim, K., Lee, S. Y., Kim, S. H., Kim, J. Y., ... & Park, J. U. 2016. Photo-patternable and transparent films using cellulose nanofibers for stretchable origami electronics. *NPG Asia Materials*, 8(8): e299.
14. Jiang, T., Huang, R., & Zhu, Y. 2014. Interfacial sliding and buckling of monolayer graphene on a stretchable substrate. *Advanced Functional Materials*, 24(3), 396–402.
15. Keplinger, C., Sun, J. Y., Foo, C. C., Rothemund, P., Whitesides, G. M., & Suo, Z. 2013. Stretchable, transparent, ionic conductors. *Science*, 341(6149), 984–987.
16. Khan, S. A., Gao, M., Zhu, Y., Yan, Z., & Lin, Y. 2017. MWCNTs based flexible and stretchable strain sensors. *Journal of Semiconductors*, 38(5): 053003.
17. Kim, S. J., Choi, K., Lee, B., Kim, Y., & Hong, B. H. 2015. Materials for flexible, stretchable electronics: Graphene and 2D materials. *Annual Review of Materials Research*, 45, 63–84.
18. Kim, T., Cho, M., & Yu, K. J. 2018. Flexible and stretchable bio-integrated electronics based on carbon nanotube and graphene. *Materials*, 11(7): 1163.
19. Le, T. S. D., An, J., & Kim, Y. J. 2017. Femtosecond laser direct writing of graphene oxide film on polydimethylsiloxane (PDMS) for flexible and stretchable electronics. *2017 Pacific Rim Conference on Lasers and Electro-Optics*.

20. Lee, H., Kim, M., Kim, I., & Lee, H. 2016. Flexible and stretchable optoelectronic devices using silver nanowires and graphene. *Advanced Materials*, 28(22), 4541–4548.
21. Lee, S. K., Jang, H. Y., Jang, S., Choi, E., Hong, B. H., Lee, J., ... & Ahn, J. H. 2012. All graphene-based thin film transistors on flexible plastic substrates. *Nano Letters*, 12(7), 3472–3476.
22. Li, H., Lv, S., & Fang, Y. 2020. Bio-inspired micro/nanostructures for flexible and stretchable electronics. *Nano Research*, 13, 1244–1252.
23. Liu, X., Zhou, X., Li, Y., & Zheng, Z. 2012. Surface-grafted polymer-assisted electroless deposition of metals for flexible and stretchable electronics. *Chemistry–An Asian Journal*, 7(5), 862–870.
24. Nguyen, B. H., & Nguyen, V. H. 2016. Promising applications of graphene and graphene-based nanostructures. *Advances in Natural Sciences: Nanoscience and Nanotechnology*, 7(2): 023002.
25. Pal, S., Sarkar, D., Roy, S. S., Paul, A., & Arora, A. 2020. Design, development and analysis of a conductive fabric based flexible and stretchable strain sensor. In *IOP Conference Series: Materials Science and Engineering* (Vol. 912, No. 2). IOP Publishing.
26. Patra, S., Choudhary, R., Madhuri, R., & Sharma, P. K. 2018. Graphene-based portable, flexible, and wearable sensing platforms: An emerging trend for health care and biomedical surveillance. In A. Tiwari, Ed., *Graphene Bioelectronics* (pp. 307–338), Elsevier.
27. Qin, J., Lan, L., Chen, S., Huang, F., Shi, H., Chen, W., ... & Yang, C. 2020. Recent progress in flexible and stretchable organic solar cells. *Advanced Functional Materials*, 30(36): 2002529.
28. Seol, Y. G., Trung, T. Q., Yoon, O.-J., & Lee, N.-E. 2012. Nanocomposites of reduced graphene oxide nanosheets and conducting polymer for stretchable transparent conducting electrodes. *Journal of Materials Chemistry*, 22(45), 23759–23766.
29. Shi, Y., Wang, X., Luo, J., & Xie, Q. 2019. Fabrication and characterization of polyoxometalate/2D graphene-based flexible supercapacitors for wearable electronic pulse-beat application. *Journal of Materials Science: Materials in Electronics*, 30(4), 3692–3700.
30. Singh, E., Meyyappan, M., & Nalwa, H. S. 2017. Flexible graphene-based wearable gas and chemical sensors. *ACS Applied Materials & Interfaces*, 9(40), 34544–34586.
31. Someya, T., Ed. (2012). *Stretchable Electronics*. John Wiley & Sons.
32. Song, J. K., Do, K., Koo, J. H., Son, D., & Kim, D. H. 2019. Nanomaterials-based flexible and stretchable bioelectronics. *MRS Bulletin*, 44(8), 643–656.
33. Sun, D. M., Liu, C., Ren, W. C., & Cheng, H. M. 2013. A review of carbon nanotube-and graphene-based flexible thin-film transistors. *Small*, 9(8), 1188–1205.
34. Uz, M., Jackson, K., Donta, M. S., Jung, J., Lentner, M. T., Hondred, J. A., & Mallapragada, S. K. 2019. Fabrication of high-resolution graphene-based flexible electronics via polymer casting. *Scientific Reports*, 9(1), 1–11.
35a. Wang, H., Wang, Z., Yang, J., Xu, C., Zhang, Q., & Peng, Z. 2018. Ionic gels and their applications in stretchable electronics. *Macromolecular Rapid Communications*, 39(16): 1800246.
35b. Wang, Y., Yang, R., Shi, Z., Zhang, L., Shi, D., Wang, E., & Zhang, G. 2011. Super-elastic graphene ripples for flexible strain sensors. *ACS Nano*, 5(5), 3645–3650.
36. Wu, D. Y., & Shao, J. J. 2020. Graphene-based flexible all-solid-state supercapacitors. *Materials Chemistry Frontiers*, 5, 557–583.
37. Xie, Z., Avila, R., Huang, Y., & Rogers, J. A. 2020. Flexible and stretchable antennas for biointegrated electronics. *Advanced Materials*, 32(15): 1902767.

38a. Yan, C., Cho, J. H., & Ahn, J. H. 2012. Graphene-based flexible and stretchable thin film transistors. *Nanoscale*, 4(16), 4870–4882.

38b. Yan, C., Wang, J., Kang, W., Cui, M., Wang, X., Foo, C. Y., ... & Lee, P. S. 2014. Highly stretchable piezoresistive graphene–nanocellulose nanopaper for strain sensors. *Advanced Materials*, 26(13), 2022–2027.

39. Yao, S., Ren, P., Song, R., Liu, Y., Huang, Q., Dong, J., ... & Zhu, Y. 2010. Nanomaterial-enabled flexible and stretchable sensing systems: Processing, integration, and applications. *Advanced Materials*, 32(15): 1902343.

40. You, R., Liu, Y. Q., Hao, Y. L., Han, D. D., Zhang, Y. L., & You, Z. 2020. Laser fabrication of graphene-based flexible electronics. *Advanced Materials*, *32*(15): 1901981.

41. Yu, J., Wu, J., Wang, H., Zhou, A., Huang, C., Bai, H., & Li, L. 2016. Metallic fabrics as the current collector for high-performance graphene-based flexible solid-state supercapacitor. *ACS Applied Materials & Interfaces*, 8(7), 4724–4729.

42. Yue, G., Ma, X., Zhang, W., Li, F., Wu, J., & Li, G. 2015. A highly efficient flexible dye-sensitized solar cell based on nickel sulfide/platinum/titanium counter electrode. *Nanoscale Research Letters* 10(1): 1.

43. Yun, S., Niu, X., Yu, Z., Hu, W., Brochu, P., & Pei, Q. 2012. Compliant silver nanowire-polymer composite electrodes for bistable large strain actuation. *Advanced Materials*, 24(10), 1321–1327.

44. Zhang, J. 2019. Flexible and stretchable devices from 2D nanomaterials. In M. Han, X. Zhang, & H. Zhang, Eds., *Flexible and Stretchable Triboelectric Nanogenerator Devices: Toward Self-Powered Systems* (pp. 149–164), Wiley.

45. Zhao, S., Li, J., Cao, D., Zhang, G., Li, J., Li, K., ... & Wong, C. P. 2017. Recent advancements in flexible and stretchable electrodes for electromechanical sensors: Strategies, materials, and features. *ACS Applied Materials & Interfaces*, 9(14), 12147–12164.

3 Elastomeric Substrate for Stretchable Electronics

Noureddine Ramdani

CONTENTS

3.1 INTRODUCTION

Stretchable electronics are an emerging type of electronics exhibiting the ability to be bent and stretched. They are produced by printing circuits and depositing these on the stretched supports. They are widely used as totally stretchable film sensors for robots, wearable electronic-based devices, wearable communication appliances, and skin-like biomachines. Researchers working on these devices are mainly focusing on how to keep their full performance and structural integrity under strong exterior stretching strains without sacrificing their mechanical robustness and electronic characteristics.

The progress and outstanding properties of smart wearable sensing devices for use in practical electronic applications have been reviewed [1].

The current approach for producing stretchable electronic devices is to incorporate rigid semiconductors into a stretchable elastomeric support. Heat management is a critical issue in the fabrication of these electronic materials as hostile thermal heat is generated, which (a) affects the materials' properties and (b) is not suitable in applications that come into contact with the human body or other organic flesh, where a rise in temperature of 1°C to 2°C is not permitted [2]. Stretchable electronics based on elastomeric substrate can demonstrate performance comparable to the usual wafer-based devices that generally exhibit stretching like a rubber band, exhibit twisting similar to a rope, can be bent easily over a pen, and can be folded as a part of sheet.

Studies on stretchable electronics mainly consider innovation of material synthesis, mechanical architecture, and production techniques based on elastomeric non-rigid substrates. The most pressing problem is that the whole device might bend and stretch as well. Thus, the elaboration of stretchable conductors is critical to ensure the connection of working circuits for these systems. In a review published by Wu [3], technological advances in fabrication of materials for stretchable electronics were explored along with the different applications of stretchable electronics. Wu also covered the shape of conductive fillers on the percolation threshold and the production techniques for fabricating elastic conductive composites: that is, inserting conductive fillers into elastomers, introducing metal fillers into an elastomeric matrix, filling microchannels with liquid metals, putting elastomers into the conductive filler morphological voids, and blending elastic polymers with some types of conductive reinforcing agents.

This chapter summarizes the various materials used as elastomeric substrate for stretchable electronics, focusing on the synthesis and fabrication of these systems as well as their various applications. In addition, there is a brief overview of technical challenges and possible future research on stretchable electronic systems based on elastomeric substrate.

3.2 TYPES OF CONDUCTIVE MATERIALS USED AS ELASTOMERIC SUBSTRATE FOR STRETCHABLE ELECTRONICS

There are three main types of conductive materials used as electrodes in stretchable devices. The first type is conventional inorganic metals including aluminum, gold, molybdenum, titanium, silver, and metal oxide. The second type uses various organic conductive polymers, including polyaniline, PEDOT:PSS, and polypyrrole. The third type uses carbon-based nanofillers like carbon nanotube and graphene sand. The fabrication of flexible, stretchable electronic devices based on hard metals and semiconductors is challenging due to the inferior strain values (cannot exceed 1%) of their elastomeric supports. Even under sustainable strains (below 5% of conductive polymers; higher than rigid metals), they cannot produce performance electronics exhibiting high stretchability. In addition, conductive polymers suffer from problems of aging over time.

3.3 PROCESSING TECHNIQUES OF ELASTOMERIC SUBSTRATE FOR WEARABLE ELECTRONICS

Various techniques have been used for fabricating nanocomposites for stretchable electronic devices, based on adding conducting/semiconducting fillers into or over an elastomeric backbone. These techniques include alloying electrical conductive fillers into some soft rubber matrices, constructing double-layered structures composed of dissimilar electronic cover and elastomer substrate film, and tailoring polymeric molecular arrangements for imparting stretchability [4]. Recently, Yu et al. summarized the development of several features of these attractive and interesting materials, which have been used as elastomeric and electrical fillers for substrate, as well as the new architectures and mechanics of stretching that reinforce modelling techniques [5]. In addition, it is suggested that given the progress with stretchable rubbery electronics, which are based on completely stretchable elastomers substrates, these could be one of the most promising technical solutions [6]. In this section, the three main approaches used to develop flexible, stretchable electronic materials are presented: wavy structural configuration, stretchable interconnects, and multidirectional writing.

3.3.1 Wavy Structural Configuration

The first method is wavy structural configuration. Rigid semiconductors, such as nanotapes, nanoropes, and nanotissues, are classified as "wavy" forms, which can bear strongly exerted strains to avoid deterioration of devices. The method used presently is based on a wavy fastening process, common in structures consisting of a tinny, rigid coating on an elastomeric substrate. As a hard, thin layer deposited on an elastomeric support is exposed to compression pressure, the thin coating alleviates superficial strain using a mechanical clipping mechanism. In the stretching mode, external strains are provided by releasing the prestrain while varying the buckles' wavelengths and amplitudes simultaneously; this significantly improves the stretching ability of the active structures or the passive interconnects. Another type of wavy structure, known as 2D horseshoe metals or 3D metal interconnects, can successfully tolerate twice the maximum elongation of the original length. Recently, Cantarella et al. reported on a new procedure to produce arbitrarily oriented and modified wrinkles on the surface of a biocompatible elastomeric support; they produced a tinny film transistor on serval circuits using an AmorphousIGZO [7].

3.3.2 Stretchable Interconnects

The second approach is based on stretchable interconnects, which interconnect stiff, dynamic, tinny device islands on elastomeric substrates using stretchable electrodes. The main technique for preparing stretched interconnects is to alloy rubber-like elastomers and highly conductive inorganic fillers. This approach is preferable as both performant electronic constituents can be successfully deposited on elastomeric supports by traditional techniques. Concerning stretchable composite interconnects, various electrically conductive fillers like metallic ropes, conductive polymers, carbon nanotubes, and graphene flakes could be used. Stretchable interconnects can be deposited on various elastomeric substrates via several techniques, such as transfer printing and screen printing, photolithographic patterning, and vacuum evaporation. However, the stretchable interconnects strategy is only useful for producing 2D structures with small aspect ratio which can be borne on supports. On the other hand, omnidirectional and discrete incorporation in stretchable electronic devices are still problems for this strategy and wavy structural configuration. Designers of stretchable circuits usually seek to orient stretchable interconnects in any direction.

Pan et al. theoretically and experimentally evaluated the effect of elastomeric substrate type on the stretching capabilities of twisting interconnects [8]. A marked enhancement in elastic stretchability due to reducing substrate thickness was revealed by finite element analysis. Thus, repeatedly wrinkled hard thin layers deposited on

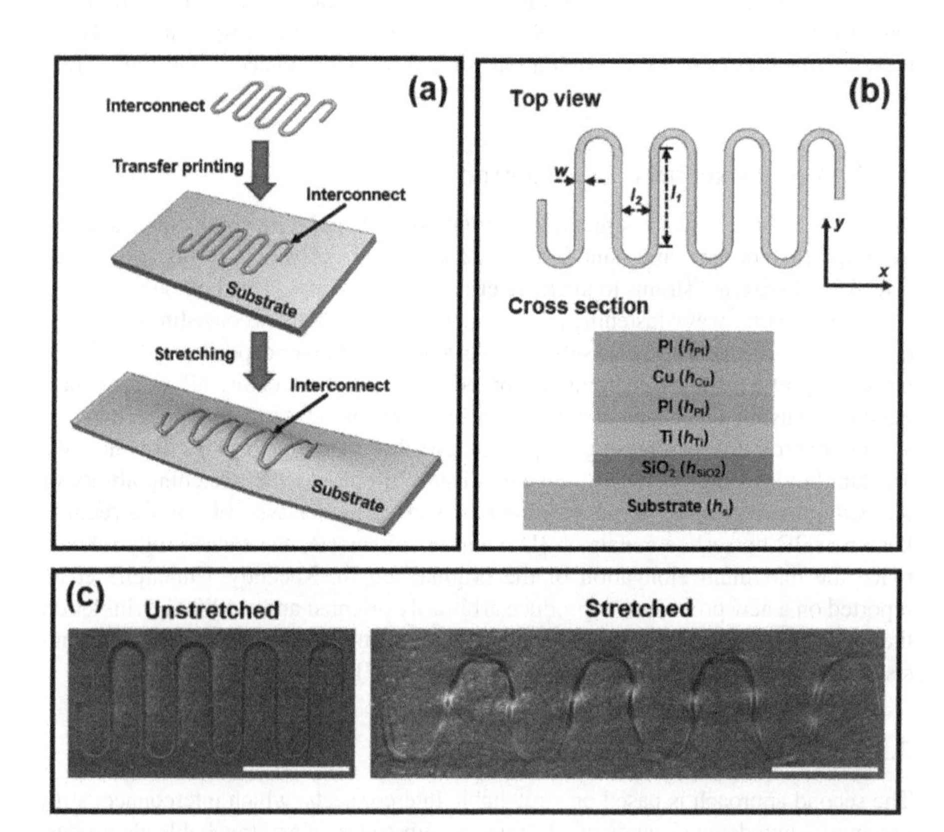

FIGURE 3.1 (a) Illustration of the steps for fabricating a serpentine interconnect on a soft substrate. (b) Geometric layout of the serpentine interconnect. (c) SEM images of the serpentines on a 500 μm-thick Ecoflex substrate in the unstretched state (left) and at 100% applied strain (right). The scale bar represents 500 μm.

Source: Reprinted with permission from [8].

soft substrates are extensively used in stretchable electronics. Figure 3.1 shows the stages for constructing a serpentine interconnect deposited on a 500-μm thick Ecoflex support – including architecture designs as well as morphologies, based on scanning electron microscopy (SEM) – in both the initial case and at 100% excerted strain.

Integration of deformable serpentine interconnects and mineral fillers with rubbery elastomeric supports provide strong deforming ability for the electronic devices, allowing randomly targeted forms. Zhao et al. proposed a new integrated material incorporating saw-like substrate serpentine interconnects exhibiting superior stretchability to release fragments of the twisting interconnects from the soft support [9].

3.3.3 Multidirectional Writing

To overcome the challenges of wavy structural configuration and stretchable interconnects, an alternative process has been widely investigated: multidirectional writing at lower temperatures of organic or mineral fillers exhibiting high electrical conductivity on soft substrates. This approach gives improved stretchability with accurately tailored forms, and the stretchable 3D assemblies can be used to develop a wide range of stretchable electronic systems and circuits on rubbery supports. Using spatioselective ultraviolet exposure to elaborate the stretchable devices with a stretchability value exceeding 100%, Cai et al. produced a soft strain-isolating polymer substrate that achieves controllable toughness [10]. Their data revealed that this approach has good potential not only for large-scale production of stretchable electronics, but also for manufacturing stretchable electronics, including the combination of multifunctional constituents.

3.4 ELASTOMERIC MATERIALS SUBSTRATES

3.4.1 Polydimethylsiloxane Substrate

The polydimethylsiloxane (PDMS) elastomer exhibits several striking characteristics including superior oxygen gas and water vapor permeability, which is critical for stretchable electronics based on organic semiconductors and oxidizable metals. Sun et al. fabricated a soft, water-resistant, and sensitive strain sensor by incorporating carbon black, graphene, and PDMS elastomer in a 3D framework of commercial Ni sponge using an easy drop-coating technique [11]. The resulting strain sensor demonstrated improved flexibility, increased sensitivity, good long-term stability, and waterproofness.

A new type of skin-like stress sensor with superior sensitivity performance was fabricated by Wang et al. from clear carbonized silk nanofiber skins and amorphous PDMS films following an economical technique [12]. Kim et al. introduced a new solution-treated type of sensor and electronic devices exhibiting rubber-like and inherently stretchable features resulting from the rubbery substrate in percolated compound arrangements with poly(3-hexylthiophene-2,5-diyl) nanofibrils and gold nanoparticles with coated silver nanowires in a PDMS matrix [13]. Characterization results showed that due to crystallinity enhancement, the resulting thin transistors maintained their electrical performance under stretching and showed one of the highest P3HT-based field-effect mobilities (1.4 $cm^2/V \cdot s$).

Sun et al. [14] achieved a low-cost, easy technique for producing ultra-stable and supper-sensitive piezoresistive sensors, based on an effective new graphene-reinforced PDMS composite, for biodevices for sensing acoustic trembling, human body movement, and radial vein beats. Kim et al. proposed a simple approach for producing printable and highly stretchable 30 μm-thick conductors by transferring printed Ag ink onto elastomeric Ecoflex substrates comprising tough hydrogel layers using a water-soluble tape [15]. The resulting conductor on hybrid film was able to be stretched to 1,780% strain, which makes it useful for wearable devices that are more comfortable for human skin. Vohra et al. reported on a method that fused a skin of PDMS elastomer onto a transparent shell of low gas permeability butyl rubber (T-IIR) support using organosilane-based adhesive, to be used in producing robust stretchable devices [16]. The most significant finding of their work was the ability for the elastomeric blend boundary to bury several microns under the PDMS phase, which significantly improved stretching performance and conductivity.

Sang et al. [17] designed a very sensitive and stretchable pressure device by adding a mixture of Ag nanowhiskers and nanoparticles into the PDMS matrix. The Ag nanoparticles enhanced conductivity by reducing the resistance of the produced sensor to 14.9 Ω. The strain sensor demonstrated a strong piezoresistivity with tailorable gauge factors at 3,766. Siloxane rubbers exhibiting improved stretchability, rapid and effective self-healing capabilities at ambient temperature, and superior mechanical performance have broadened their prospects for use in several areas. Zhao et al. prepared a modified type of siloxane rubber by incorporating aromatic disulfides into a siloxane host [18]. At room temperature, the newly developed material demonstrated good tensile stress, very high elongation strain, and improved self-healing capability.

3.4.2 Block Copolymer Elastomers

Block copolymer (BC) elastomers play a very important part in the fabrication of wearable electronics. Because BCs are supported by their physical cross-linking forms, this makes them soft elastomers. Their physical characteristics are different from ordinary elastomers produced using a chemical cross-linking process. They do not require an additional chemical cross-linking treatment; thus they are cast directly and subsequently melted in different solvents. The good viscoelastic and thermoplastic features of BCs make them easily moldable and sticky. Their striking physical characteristics can be considered drawbacks in some applications, but they are dynamically useful in producing many types of stretchable electronic devices, and it is expected that their use will increase in the future.

After the deposition process, conductive fillers, like nanowires or nanosheets, are preferred at higher aspect ratios so as to be infiltrated well by the BCs. During the infiltration stage, the contacts between the fillers are kept. This makes tinny hybrid films exhibiting improved stretchability and higher conductivity. The metal precursor solution printing technique is considered a powerful processing method due to its compatibility with old printing methods, with no blockage of jets, and its capacity for high reinforcement loading. Using a BC elastomer like a support, it has been found that a double layer of BC/PDMS can be formed due to their good viscoelastic and thermoplastic properties. The double-layer substrate can exhibit a perfect elastomeric comportment, and the benefits of the BC support can be preserved in the case of a BC/PDMS double layer having a thicker PDMS layer, as opposed to using a viscoelastic BC monosystem. In addition, the use of traditional fabricating methods can be critical for marketing the various stretchable electronic devices. BC supports exhibiting microfibril linkage on their exteriors ease the manufacture of high-tenacity circuits by placing metals over a cover on the support. Some work on elaborating stretchable biological transistors has been recorded based on in situ phase departure of polymeric semiconductors for forming nanofibril superficial packages of a BC. In this way, a high-resolution transistor collection with large surface area was produced. The applicability of BC elastomers has grown in areas like stretchable batteries, where there have been many feasibility studies on development of different nanocomposites, pore size-organized skins, and micro-arranged surfaces. Although, for use of these BCs in the production of wearable biomedical devices, their long-term stability under heat, solvents, and ultraviolet radiation must be guaranteed.

To improve mechanical performance and sensor sensitivity to human skin, Hanif et al. elaborated an ultrathin membrane-like support showing high stretchability, good transparency, and high toughness by incorporating piezoelectric poly((vinylidene fluoride)-co-trifluoroethylene) (P-(VDF-TrFE)) nanofibers exhibiting a high modulus into PDMS elastomeric having low modulus [19]. The resulting skin-like substrate was covered with a stretchable temperature sensor and demonstrated strong ability to accommodate body motion-induced strain without sacrificing its inherent mechanosensory and thermosensory features.

You et al. reviewed the latest usage aspects of BC elastomers in stretchable electronic materials and discussed future applications as well as problems to be overcome for everyday practice [20]. Research on these kinds of materials initially focused on the production technique for stretchable electrodes, the physical alloying of electrically conductive fillers and BCs, the penetration of BCs into a conductive film, and finally the conversion of organometallics into metallic nanofillers within the BC substrate.

3.4.3 Polyurethane Substrate

Polyurethane (PU) is one of the most important types of polymeric elastomers. It is constituted from a periodic structure of carbonate chain between rubber and plastic. It is the only polymer exhibiting numerous outstanding physical and chemical characteristics. These features are chemically controlled throughout the preparation process or physically controlled using certain stabilizers.

Wang et al. explored the technique of combining an elastomer nanofiber membrane and metallic conductive dopant as a cost-effective method for large-scale production of capacitive strain sensors [21]. They developed an effective capacitive pressure sensor from a stretchable support based on carbon nanotubes with painted PU rubber skins and silver nanowires as a conductor. This was constructed to perceive small body motions.

Wang et al. also developed a new type of elastomeric polyurethane reinforced with silver nanowire-based nanocomposite network conductor which exhibits several advantageous properties, such as low mechanical defiance, improved strain-invariant conductance, and enhanced breathability [22]. The developed nanocomposite mesh conductor is useful for producing epidermal microelectronic devices.

In a study conducted by Li et al., very robust ionic membranes demonstrating excellent healing features and enhanced sensitivity were elaborated by including ionic fluids in a tough poly(urea-urethane) network, which was synthesized from poly(ε-caprolactone) and poly(ethylene glycol) biopolymers [23]. Their resulting ionic membranes exhibited high mechanical performance, high sensitivity, improved elasticity, and outstanding healability. Xu et al. [24] produced a reinforced-polyurethane filled with graphite nanoplatelet as a new capacitive-type pressure sensor using a cost-effective gap coating process. These sensors were easily stretched to 30% and demonstrated a remarkable negative gauge factor of ~3.5. Figure 3.2 demonstrates the various characteristics of the developed skin, including: self-healing effect,

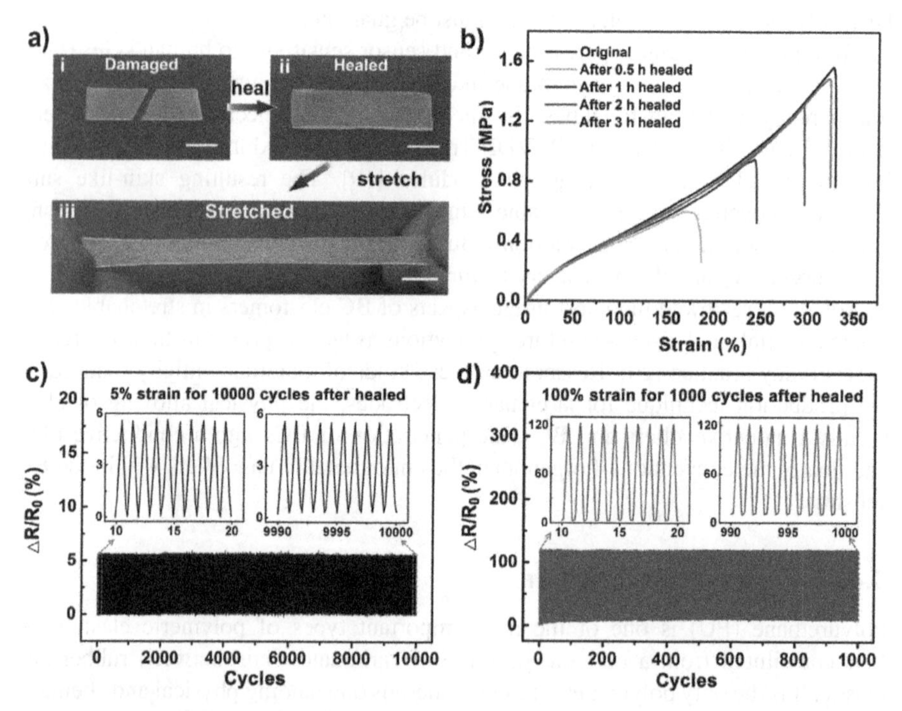

FIGURE 3.2 (a) Digital images of PU-IL2 ionogel (i) cut into two halves, (ii) after healing for 2 hours, and (iii) after stretching. The scale bar is 1 cm. (b) Stress–strain curves of pristine PU-IL2 ionogel and PU-IL2 ionogel previously cut into two halves and healed for 0.5, 1, 2, and 3 hours. Cyclic stability tests of the healed I-skin under (c) 5% strain for 10,000 cycles and under (d) 100% strain for 1,000 cycles. The insets are the enlarged signals of relative resistance changes from the 11th to 20th cycle and over the last 10 cycles.

Source: Reprinted with permission from [23]

stress–strain curves, and cyclic stability experiments of the rehabilitated I-skin below 5% deformation for 10^4 cycles and under 100% strain for 10^3 cycles.

3.4.4 Ionic Gels Elastomeric Substrates

Currently, the most affordable stretchable electronics consist of electrically conductive materials deposited on soft elastomeric membranes. Ionic gels consist of dielectric elastomeric nets and electrolytic solutions-based stretchable materials. These are summarized by Wang et al. [25], who focus on the ionic gel architectures, the manufacturing methods of ionic-gel-based stretchable electronic devices, and their practical utilization. Several elastomers, including rubber, polyurethanes, silicones, and acrylic elastomers, are representative substrates used in producing dielectric elastomeric sensors. Wang et al. produced a conductive-active clear synthetic muscle using self-directed regenerated ionic-gel-based conductors [26]. By connecting ion–dipole contacts between the fluorinated rubber and the ionic liquids, the resulting gels exhibited advantageous self-restorative properties while maintaining their mechanical performance, which was totally reinstated within one day, showing that the lifetime of these actuators was improved.

3.5 APPLICATIONS OF ELASTOMERIC SUBSTRATE-BASED WEARABLE ELECTRONICS

Those researching wearable electronics are continuously seeking to develop conductive rubber substrates that combine good mechanical stretchability and improved electrical conductivity for practical uses [27]. There has been an increasing need for wearable devices in recent years. This includes stretchable, membrane-mountable, human–machine interfaces, energy harvesting, wearable strain sensors, tailored health monitoring, human signs detection, soft robotics, and many others. In this section, the various applications of elastomeric substrate-based stretchable and wearable electronics is presented in detail with examples from the literature.

3.5.1 Flexible Strain Sensors

One well-known smart wearable device is the flexible pressure sensor. Jian et al. produced high-performance strain sensors based on biomimetic microstructured PDMS architectures and hybrid highly conductive carbon nanotubes and graphene-based active skins. The device they produced was suitable for use as a human–machine interface or in wearable electronics [28]. Another type of strain sensor based on elastomeric semiconductors that exploits π-π casted poly(3-hexylthiophene-2,5-diyl) nanofibrils infiltrated into PDMS was produced by Kim et al. [29]. This sensor showed a linear response, an increased stretchability, great tolerance for mechanical deformations, and only low hysteresis. Liu et al. fabricated a wearable multifunctional strain sensor form an Ecoflex substrate doped with graphene and a new ionic conductor as the sensing film [30]. The strain sensor can be stretched to 300% under the different mechanical strains. Liu et al. also produced a stable strain sensor by inserting reduced graphene oxide-covered polystyrene microspheres and nanowires of conductive silver into a viscoelastic PDMS [31]. Similarly, Wu et al. developed a highly sensitive strain sensor from a graphene aerogel and PDMS to tailor its sensitivity by adjusting the hexagonal architecture; they achieved various fabrication techniques [32]. Chen et al. fabricated a multifunctional strain sensor by polymerizing pyrrole within a cross-linked polymeric matrix [33]. The elaborated strain sensor was suitable for detecting different human motions. An inherently stretchable transistor consisting of single-walled carbon nanotube probes and a non-polar elastomer was described by Chortos et al. [34]. The produced material exhibited high sensitivity to strain due to a marked dependency on its dielectric's depth. Azizkhani et al. presented a strain sensor demonstrating high stretchability and sensitivity with economical processing, fabricated by adding sliced carbon fibers inserted into elastomeric silicone elastomeric films [35].

Jiang et al. studied a various strain distribution of several elastomeric substrates to considerably improve the sensitivity of stretchable strain sensors [36]. Lee et al. introduced a new, simple structural engineering approach consisting of a soft membrane and a microtrench-patterned elastic support to elaborate a new generation of strain sensors [37]. Figure 3.3(a–i) shows some details of the performance and usage mode of the developed sensor.

3.5.2 Biomedical Applications

In the last few years, remote and personalized health monitoring based on elastomeric-based stretchable electronic sensors has become a prominent solution to support medical diagnosis and ensure timely treatment. Table 3.1 shows the possible applications of remote and personalized health monitoring based on wearable sensors that gather various physiological data from the human body. Gao et al. reviewed recently developed sensitive and stretchable sensors using new architectures or new materials [38].

To facilitate personal health monitoring, the materials used for wearable devices should be lightweight, flexible, and tinny. In this context, Gong et al. produced latex elastomer containing very elastic dark gold e-skin nanopatches as a highly sensitive biomedical sensor [39]. Yang et al. reviewed the applications of wearable sensors in chronic disease care and carried out a comprehensive investigation into requirements for flexibility and stretchability and methods of nano-based improvement [40]. Xi et al. developed a microtubular sensor exhibiting high sensitivity to mechanical deformations, with the sensor forming a liquid core within a flexible silicone rubber microtube [41]. The application of this microtubular sensor to record arterial pulse at the wrist above a period of 10 seconds before and after exercise is illustrated in Figure 3.4.

3.5.3 Human–Machine Interfaces, Soft Robotics, and Haptics

Wearable electronics are principally designed to improve human–machine interfacing. By using suitable strain sensors, the actuation of smart robots can be achieved. Another use for elastomeric-based wearable electronics can be found in smart gloves. Recently, Amjadi et al. [42] proposed a new type of strain sensor with graphite layers covering an elastomer film; this displayed highly sensitivity and stretchability, and

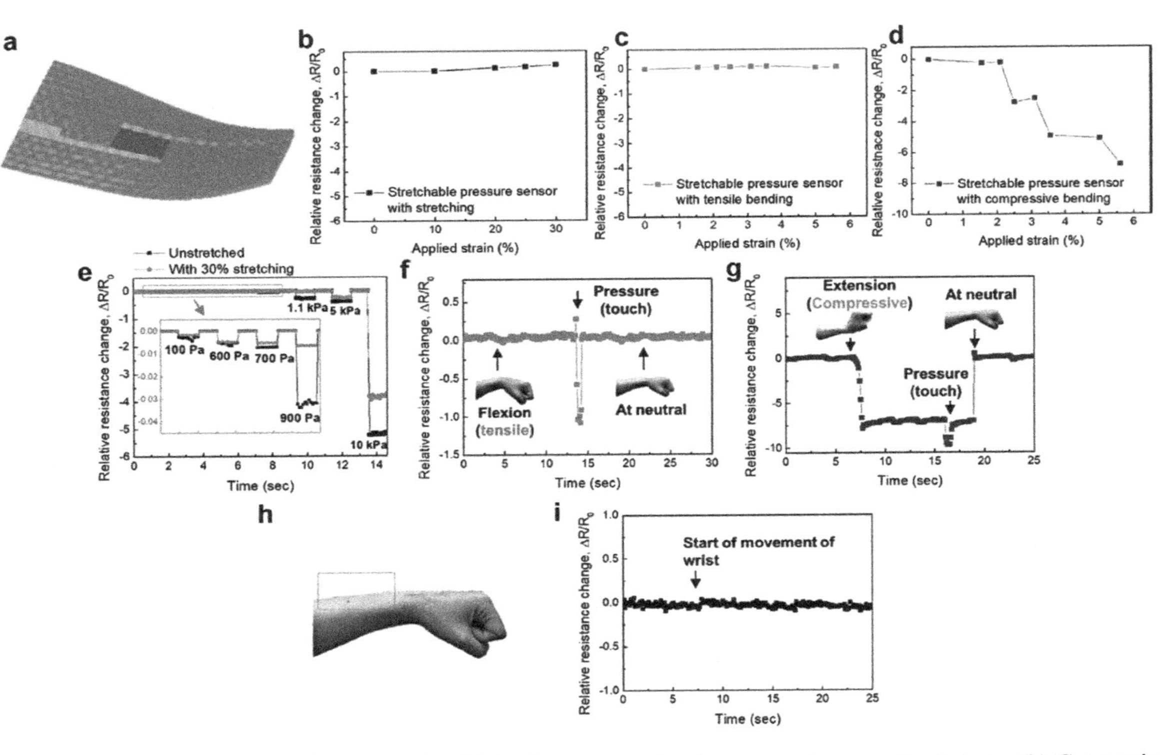

FIGURE 3.3 (a) Illustration of the steps for fabricating a serpentine interconnect on a soft substrate. (b) Geometric layout of the serpentine interconnect. (c) SEM images of the serpentines on a 500 μm-thick Ecoflex substrate in the unstretched state (left) and at 100% applied strain (right). The scale bar represents 500 μm.

Source: Reprinted with permission from [8]

TABLE 3.1
Biomedical applications of elastomeric substrate-based stretchable sensors

Type of measure	Location	Sensor type	Elastomeric substrate	Challenges
Body motion	Joint, wrist, spine, face, chest, neck	Resistance Tremors Accelerometer sensor	Silicon nanomembrane	Large-scale stretchability, fast response/recovery speeds, sensitivity
Temperature	Whole body	Pyroelectric temperature detectors	PDMS-CNT/P(VDFTrFE)/graphene	Response and relaxation times
Respiration rate	Nose, mouth, chest	Tunneling piezoresistance	CNT/PDMS	Esthetics, comfort
		Volume sensor	PPy/PU	
Blood pressure	Wrist, neck	Piezoresistance	PEDOT:PSS/PUD/PDMS	Sensitivity

can be applied in soft robotics, human physical motion appreciation, vibration recognition, and human body large motion capturing.

3.5.4 Energy Storage and Harvesting Devices

In recent years, extensive research has been conducted on the development of independent energy harvesters with highly stretchable electronic systems, to be used as power sources. The development of highly flexible and stretchable energy harvesters, those exhibiting a stretchability strain exceeding 15%, was reviewed by Wu et al. [43]. They mainly focused on approaches to elaboration of materials, device fabrication, and types of integration for enhancing both stretchability and flexibility. Jinno et al. developed very flexible carbon-based photovoltaics covered on both laterals with elastomeric material to ensure stretchability and stability that retains its efficiency when in water [44].

By using a liquid iron alloy as an anode, a carbon adhesive, MnO_2 as a cathode, a basic electrolytic hydrogel, and a soft elastomeric package, Liu et al. successfully fabricated a rechargeable, flexible battery [45]. Figure 3.5 shows the battery-powered strain sensor mounted on the wrist and the strain sensor extends during stretching with a frequency of ≈1 Hz. This battery included into a wearable elastomeric cover powered a blue light-emitting diode (LED) and a strain-sensing circuit (Figure 3.5 (a)).

3.6 CHALLENGES AND OUTLOOK

The development of elastomeric substrate for stretchable electronic devices mainly focuses on a single elastomer – PDMS. The design of wearable electronics with good flexibility and robustness based on such elastomers is recommended in the future

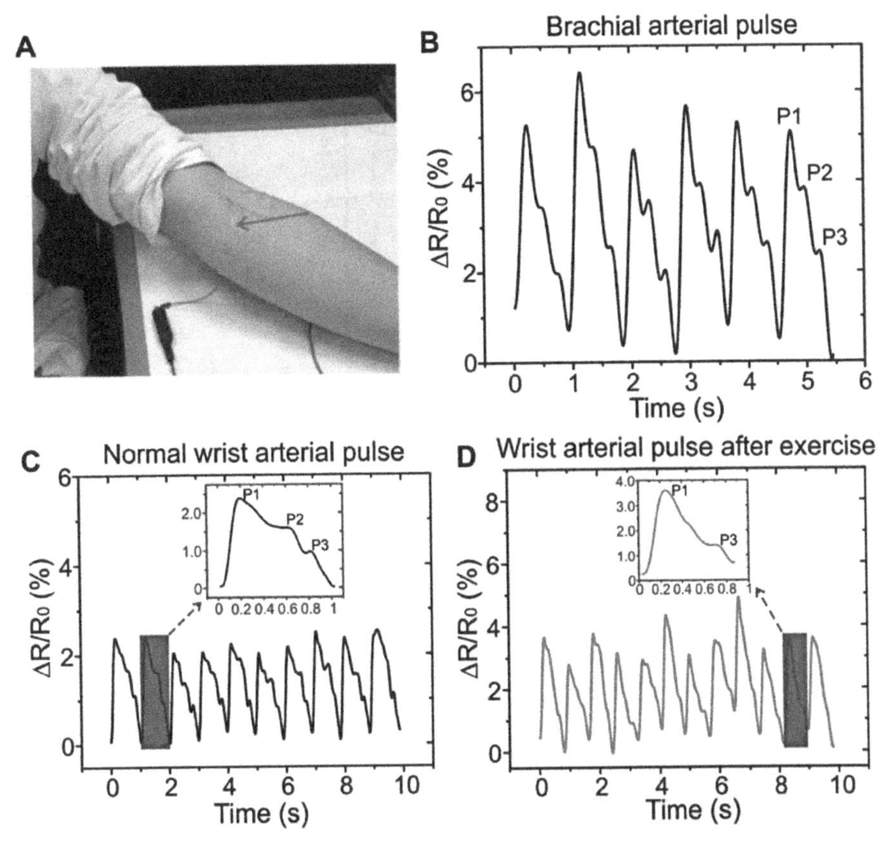

FIGURE 3.4 Applications of the liquid-based microtubular sensor. (a) Photograph showing the sensor attached firmly over the brachial artery at the elbow. The blue arrow indicates the position of the sensor. (b) Relative electrical resistance change (ΔR/R0) of the tactile sensor reflecting the brachial arterial pulse waveform. P1, P2, and P3 denote three distinct peaks indicative of the incident, tidal, and diastolic wave. (c) Sensor recording at the wrist over a period of 10 seconds. Inset shows the representation of the wrist arterial pulse waveform. (d) Sensor recording at the wrist exercise. Inset shows the representation of the wrist arterial pulse waveform, with P1 and P3 peaks.

to provide mechanical performance, to give a lenient surface for interrelating with human users, and to provide an improved physical barrier to protect the stretchable devices from the hostile atmosphere. Although, it is a real problem for conventional rubbers to meet all these characteristics, novel research on the synthesis of new elastomers and the modification of their structures with nanomaterials are appearing as promising solutions. Thus, a continuing challenge in producing skin-attachable stretchable devices is the formation of concentrated local stress on the coatings of these stretchable electronics. These stretchable electronics are dependent on mechanical strain generated by gestures, provoking cracks due to strain-persuaded signal interference phenomena.

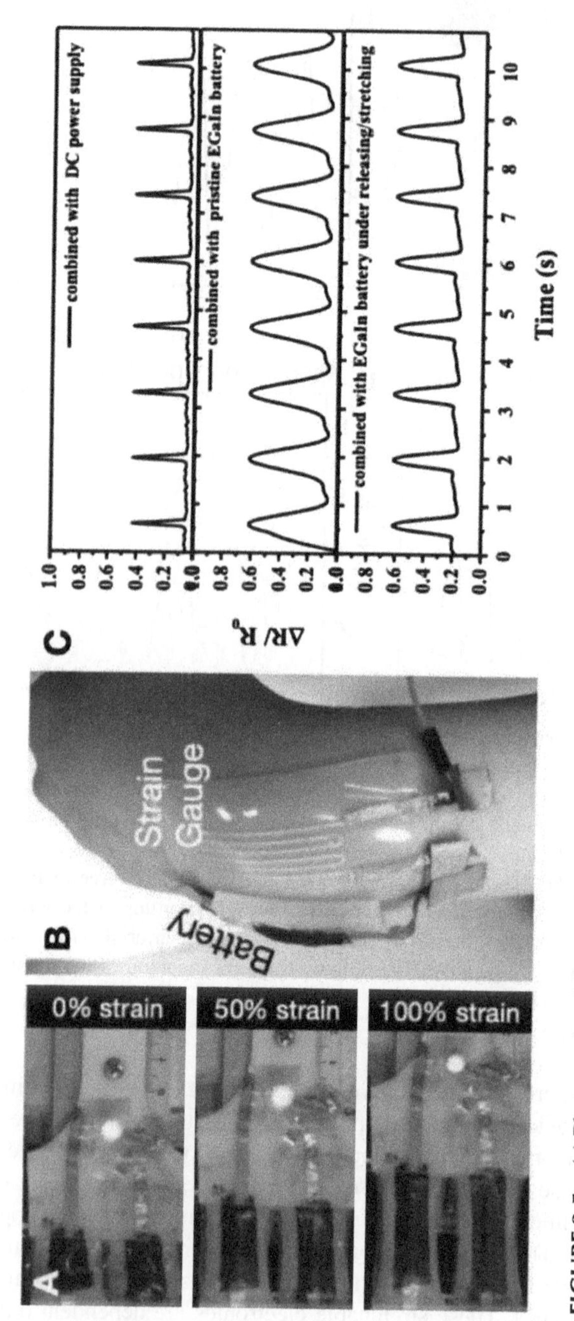

FIGURE 3.5 (a) Photographs of the stretchable EGaIn-MnO_2 battery array in series of two stretched under 0%, 50%, and 100% strain integrated with LEDs. (b) Photograph of battery-powered strain sensor mounted on the wrist. (c) Strain sensor measurements during stretching with a frequency of ≈1 Hz when powered with a DC power supply, pristine EGaIn battery, and releasing/stretching EGaIn battery, respectively.

Source: Reprinted with permission from [45]

3.7 CONCLUSIONS

This chapter reviewed recent developments in the synthesis and applications of the wearable electronic systems based on elastomer substrates in various architectures. The extensive studies on structures, elastomeric substrates, fabrication techniques, and their integrations for producing various high-performance stretchable electronic devices were summarized. Optimal conductive elastomers have been mainly produced from various groupings of non-conductive elastomers and rigid conductive fillers. According to the elastomeric material combinations that have been used and the resulting structures, steerable electronics can be utilized in a specific area. To achieve further development of stretchable technology, several challenges have to be overcome, such as ease of material assembly and device architecture, guarantee of both reproducible and reliable characteristics, and simplification of preparation approaches.

REFERENCES

1. Lou, Z., Wang, L., and Shen, G. 2018. Recent advances in smart wearable sensing systems. *Advanced Materials Technology*, 3(12): 1800444.
2. Yu, K. J., Yan, Z., Han, M., and Rogers, J. A. 2017. Inorganic semiconducting materials for flexible and stretchable electronics. *NPJ Flexible Electronics*, 1: 4.
3. Wu, W. 2019. Stretchable electronics: Functional materials, fabrication strategies and applications. *Science and Technology of Advanced Materials*, 20(1): 187–224.
4. Kim, D. C., Shim, H. J., Lee, W., Koo, J. H., and Kim, D.-H. 2019. Material-based approaches for the fabrication of stretchable electronics. *Advanced Materials*, 32(15): 1902743–1902772.
5. Yu, X., Mahajan, B. K., Shou, W., and Pan, H. 2017. Materials, mechanics, and patterning techniques for elastomer-based stretchable conductors. *Micromachines*, 8(1): 7.
6. Sim, K., Rao, Z., Ershad, F., and Yu, C. 2019. Rubbery electronics fully made of stretchable elastomeric electronic materials. *Advanced Materials*, 32(15): 1902417.
7. Cantarella, G., Vogt, C., Hopf, R., Münzenrieder, N., Andrianakis, P., and Petti, L. 2017. Buckled thin-film transistors and circuits on soft elastomers for stretchable electronics. *ACS Applied Materials & Interfaces*, 9(34): 28750–28757.
8. Pan, T., Pharr, M., Ma, Y., Ning, R., Yan, Z., Xu, R. et al. 2017. Experimental and theoretical studies of serpentine interconnects on ultrathin elastomers for stretchable electronics. *Advanced Functional Materials*, 27(37): 1702589.
9. Zhao, Q., Liang, Z., Lu, B., Chen, Y., Ma, Y., and Feng, X. 2018. Toothed substrate design to improve stretchability of serpentine interconnect for stretchable electronics. *Advanced Materials Technology*, 3(11): 1800169.
10. Cai, M., Nie, S., Du, Y., Wang, C., and Song, J. 2019. Soft elastomers with programmable stiffness as strain-isolating substrates for stretchable electronics. *ACS Applied Materials & Interfaces*, 11(15): 14340–14346.
11. Sun, S., Liu, Y., Chang, X., Jiang, Y., Wang, D., Tang, C. et al. 2020. Wearable, waterproof, and highly sensitive strain sensor based on three-dimensional graphene/carbon black/Ni sponge for wirelessly monitoring human motion. *Journal of Materials Chemistry C*, 6: 2074–2085.
12. Wang, Q., Jian, M., Wang, C., and Zhang, Y. 2017. Carbonized silk nanofiber membrane for transparent and sensitive electronic skin. *Advanced Functional Materials*, 27(9): 1605657.

13. Kim, H.-J., Sim, K., Thukral, A., and Yu, C. 2017. Rubbery electronics and sensors from intrinsically stretchable elastomeric composites of semiconductors and conductors. *Science Advances*, 3(9): e1701114.
14. Sun, Q.-J., Zhuang, J., Venkatesh, S., Zhou, Y., Han, S.-T., Wu, W. et al. 2018. Highly sensitive and ultrastable skin sensors for biopressure and bioforce measurements based on hierarchical microstructure. *ACS Applied Materials & Interfaces*, 10(4): 4086–4094.
15. Kim, S. H., Jung, S., Yoon, I. S., Lee, C., Oh, Y., and Hong, J.-M. 2018. Ultrastretchable conductor fabricated on skin-like hydrogel–elastomer hybrid substrates for skin electronics. *Advanced Materials*, 30(26): 1800109–1800117.
16. Vohra, A., Schlingman, K., Carmichael, R. S., and Carmichael, T. B. 2018. Membrane-interface-elastomer structures for stretchable electronics. *Chemistry*, 4(7): 1673–1684.
17. Shengbo, S., Lihua, L., Aoqun, J., Qianqian, D., Jianlong, J., Quiang, Z. et al. 2018. Highly sensitive wearable strain sensor based on silver nanowires and nanoparticles. *Nanotechnology*, 29(25): 1–22.
18. Zhao, L., Yin, Y., Jiang, B., Guo, Z., Qu, C., and Huang, Y. 2020. Fast room-temperature self-healing siloxane elastomer for healable stretchable electronics. *Journal of Colloid and Interface Science*, 573: 105–114.
19. Hanif, A., Trung, T. Q., Siddiqui, S., Toi, P. T., and Lee, N.-E. 2018. Stretchable, transparent, tough, ultrathin, and self-limiting skinlike substrate for stretchable electronics. *ACS Applied Materials & Interfaces*, 10(32): 27297–27307.
20. You, I., Kong, M., and Jeong, U. 2019. Block copolymer elastomers for stretchable electronics. *Accounts of Chemical Research*, 52(1): 63–72.
21. Wang, J., Lou, Y., Wang, B., Sun, Q., Zhou, M., and Li, X. 2020. Highly sensitive, breathable, and flexible pressure sensor based on electrospun membrane with assistance of AgNW/TPU as composite dielectric layer. *Sensors*, 20(9): 2459.
22. Wang, J., Zhang, K., Wang, J., Zhang, M., Zhou, Y., Cheng, J. et al. 2020. Strain-invariant conductance in elastomeric nanocomposite mesh conductor for stretchable electronics. *Journal of Materials Chemistry C*, 8(27): 9440–9448.
23. Li, T., Wang, Y., Li, S., Liu, X., and Su, J. 2020. Mechanically robust, elastic, and healable ionogels for highly sensitive ultra-durable ionic skins. *Advanced Materials*, 32(32): 2002706.
24. Xu, J., Wang, H., Ma, T., Wu, Y., Xue, R., Cui, H. et al. 2020. A graphite nanoplatelet-based highly sensitive flexible strain sensor. *Carbon*, 166: 316–327.
25. Wang, H., Wang, Z., Yang, J., Xu, C., Zhang, Q., and Peng, Z. 2018. Ionic gels and their applications in stretchable electronics. *Macromolecular Rapid Communications*, 39(16): 1800246.
26. Cao, Y., Morrissey, T. G., Acome, E., Allec, S. I., Wong, B. M., Keplinger, C. et al. 2017. A transparent, self-healing, highly stretchable ionic conductor. *Advanced Materials*, 29(10): 1605099.
27. Noh, J.-S. 2016. Conductive elastomers for stretchable electronics, sensors and energy harvesters. *Polymers*, 8(4): 123–141.
28. Jian, M., Xia, K., Wang, Q., Yin, Z., Wang, H., Wang, C. et al. 2017. Flexible and highly sensitive pressure sensors based on bionic hierarchical structures. *Advanced Functional Materials*, 27(9):1606066.
29. Kim, H.-J., Thukral, A., and Yu, C. 2018. A highly sensitive and very stretchable strain sensor based on a rubbery semiconductor. *ACS Applied Materials and Interfaces*, 10(5): 5000–5006.
30. Liu, C., Han, S., Xu, H., Wu, J., and Liu, C. 2018. A multifunctional highly sensitive multiscale stretchable strain sensor based on a graphene/glycerol–KCl synergistic conductive network. *ACS Applied Materials and Interfaces*, 10(37): 31716–31724.

31. Liu, X., Liang, X., Lin, Z., Lei, Z., Xiong, Y., Hu, Y. et al. 2020. Highly sensitive and stretchable strain sensor based on a synergistic hybrid conductive network. *ACS Applied Materials Interfaces*, 12(37): 42420–42429.
32. Wu, S., Ladani, R. B., Zhang, J., Ghorbani, K., Zhang, X., Mouritz, A. P. et al. 2016. Strain sensors with adjustable sensitivity by tailoring the microstructure of graphene aerogel/PDMS nanocomposites. *ACS Applied Materials & Interfaces*, 8(37): 24853–24861.
33. Chen, J., Liu, J., Thundat, T., and Zeng, H. 2019. Polypyrrole-doped conductive supramolecular elastomer with stretchability, rapid self-healing, and adhesive property for flexible electronic sensors. *ACS Applied Materials & Interfaces*, 11(20): 18720–18729.
34. Chortos, A., Zhu, C., Oh, J. Y., Yan, X., Pochorovski, I., To, J. W. F. et al. 2017. Investigating limiting factors in stretchable all-carbon transistors for reliable stretchable electronics. *ACS Nano*, 11(8): 7925–7937.
35. Azizkhani, M. B., Rastgordani, S., Anaraki, A. P., Kadkhodapour, J., and Hadavand, B. S. 2020. Highly sensitive and stretchable strain sensors based on chopped carbon fibers sandwiched between silicone rubber layers for human motion detections. *Journal of Composite Materials*, 54(3): 1–12.
36. Jiang, Y., Liu, Z., Wang, C., and Chen, X. 2019. Heterogeneous strain distribution of elastomer substrates to enhance the sensitivity of stretchable strain sensors. *Accounts of Chemical Research*, 52(1): 82–90.
37. Lee, J., Roh, E., and Lee, N.-E. 2020. microtrench-patterned elastomeric substrate for stretchable electronics with minimal interference by bodily motion. *Advanced Materials for Technologies*, 5(8): 2000432.
38. Gao, Q., Zhang, J., Xie, Z., Omisore, O., Zhang, J., Wang, L. et al. 2019. Highly stretchable sensors for wearable biomedical applications. *Journal of Materials Science*, 54: 5187–5223.
39. Gong, S., Lai, D. T. H., Su, B., Si, K. J., Ma, Z., Yap, L. W. et al. 2015. Latex rubber highly stretchy black gold e-skin nanopatches as highly sensitive wearable biomedical sensors. *Advanced Electronic Materials*, 1(4): 1400063–1400070.
40. Yang, G., Pang, G., Pang, Z., Gu, Y., Mäntysalo, M., and Yang, H. 2018. Non-invasive flexible and stretchable wearable sensors with nano-based enhancement for chronic disease care. *IEEE Reviews In Biomedical Engineering*, 12: 34–71.
41. Xi, W., Yeo, J. C., Yu, L., Zhang, S., and Lim, C. T. 2017. An ultrathin and wearable microtubular epidermal sensor for real-time physiological pulse monitoring. *Advanced Materials Technology*, 2(5): 1700016.
42. Amjadi, M., Turan, M., Clementson, C. P., and Sitti, M. 2016. Parallel microcracks based ultrasensitive and highly stretchable strain sensors. *ACS Applied Materials Interfaces*, 8(8): 5618–5626.
43. Wu, H., Huang, Y. A., Xu, F., Duan, Y., and Yin, Z. 2016. energy harvesters for wearable and stretchable electronics: From flexibility to stretchability. *Advanced Materials*, 28(45): 9881–9919.
44. Jinno, H., Fukuda, K., Xu, X., Park, S., Suzuki, Y., Koizumi, M. et al. 2017. Stretchable and waterproof elastomer-coated organic photovoltaics for washable electronic textile applications. *Nature Energy*, 2: 780–785.
45. Liu, D., Su, L., Liao, J., Reeja-Jayan, B., and Majid, C. 2019. Rechargeable soft-matter EGaIn-MnO_2 battery for stretchable electronics. *Advanced Energy Materials*, 9(46): 1902798.

4 Highly Sensitive Long-Term Durability of Wearable Biomedical Sensors

Vinoth Rathinam, Sasireka Rajendran, and Sugumari Vallinayagam

CONTENTS

4.1 INTRODUCTION

In biology, biochips are microprocessor chips that can be widely used. Originally created in 1983 to control fisheries, the biochip system now covers more than 300 zoos and more than 80 government organizations in at least 20 countries, and it is used for pets, electronic "branding" of horses, laboratory animal surveillance, monitoring of endangered species as well as tracking clothing, cars, dangerous waste, and humans. Biochips are inching "silently" toward humans. For example, at least 6 million medical devices, such as prosthetic devices, breast implants, and chin implants, have been implanted. In particular, higher surface ratios contribute to decreased chemical requirements, improved control, reduced waste, faster processing, and immense capacity for integration of processes. Lab-on-chip technology has potential to have a significant socio-economic impact. In laboratory medicine, completely incorporated micro-level devices for chemical production and different diagnostics show promising technological breakthroughs and offer a paradigm change in chemical processing. The Human Genome Project has contributed greatly to this technological development [1, 2].

4.2 WHAT ARE BIOCHIPS?

In the most generic sense, any system or component containing biological resources, either derived from biological origin or manually synthesized in a laboratory, on a solid substrate can be considered to be a biochip. However, in practical terms, biochips must also be efficient, typically in microarray format, and cost effective to manufacture. The artificial nose chip, DNA micro-array chip, polymerase chain reaction chip, electronic tongue, protein chip and biochemical lab-on-a-chip are some examples which meet these criteria. The most complex biochip research has been performed in relation to the protein chip and gene chip. In this review, the chip is used to demonstrate the interdisciplinary nature of this new technology. Much attention has been paid to biochips combining traditional biotechnology with unique semiconductor processing, optoelectronics, micro-electro-mechanical systems (MEMS), image acquisition, and optical signal processing [3, 4].

4.3 BIOCHIP TECHNOLOGY

To drug manufacturers, the idea of cost-effective and stable chips that conduct thousands of biological processes has wide appeal. Since these biochips can carry out extremely repetitive laboratory tasks with prescribed, microfluidic assay chemistries, they can offer ultra-sensitive recognition methodologies in a considerably smaller amount of space at extensively less cost per assay than conventional methods. Currently, applications are mainly based on testing genetic materials for mutations or differences in arrangement. Corporate attention focuses on the potential of biochips to be used for drug lead detection either in point-of-care diagnostics or in high-throughput screening platforms. The main obstacle to making this industry as widely relevant in the computer trade as processor chips is the development of an identical chip platform that can stimulate widespread implementation with a variety of "motherboard" systems [5, 6].

4.4 BIOSENSORS

A biosensor is a diagnostic system that identifies changes in biological elements or materials, such as enzymes, tissues, microorganisms, cells, and acids, and transforms this into an electrical signal. It has a transducer and a biological element that transforms the data into electrical signals. There will also be an electronic circuit consisting of a processor, a signal conditioning unit or microcontroller, and a display unit. Selectivity, stability, cost, sensitivity, linearity, and reproducibility are the key characteristics of biosensors.

4.4.1 Biosensor Components

The components of a biosensor are presented in Figure 4.1. A typical biosensor will have the following parts.

- **Analyte:** A molecule of interest that is to be detected by a suitable biosensor.
- **Bioreceptor:** A substance that exclusively recognizes the analyte molecule. For example, a variety of enzymes, cells, antibodies and nucleic acids can be used. The production of a signal (in the form of a change in heat, light, charge, acidity, mass, etc.) upon communication of bioreceptors with analytes is known as "biorecognition."
- **Transducer:** The transducer is an effective element that changes one form of energy into another form. A transducer is used to change the biorecognition event into a computable signal. This process of energy translation is known as "signalization."
- **Electronics:** A section that processes and prepares the transmitted signal for display. It consists of complex electronic circuits, such as magnification and alteration of signals from analogue to digital forms that perform signal conditioning. Processed signals are quantified by the biosensor display components.
- **Display:** The display contains a system to communicate to users; for example, liquid crystal displays (LCDs) showing numbers or curves that are understandable to users. The bioreceptor includes biologically derived materials or biomimetic components that interact (bind or recognize) with a particular bioanalyte. The transducer converts the signal produced by the real bioanalyte's interaction into another signal that can be measured and quantified more easily. The transducer is often based on one or more mechanisms, such as physicochemical, optical, piezoelectric, and/or electrochemical effects. Biosensor production was first documented in the early 1960s. Currently, biosensors find applications in two main fields: biological monitoring and environmental sensing [7–10].

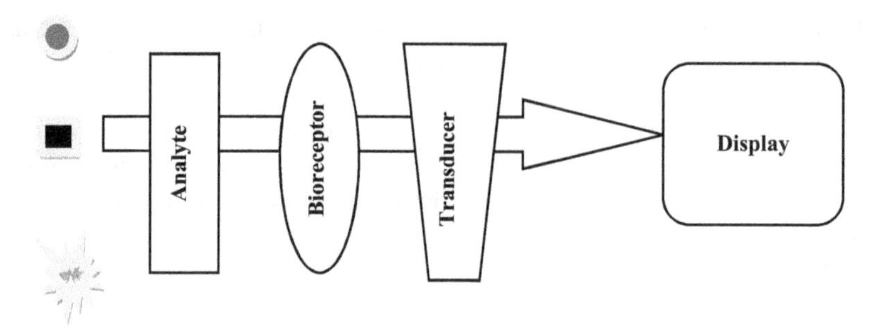

FIGURE 4.1 Schematic representation of a biosensor.

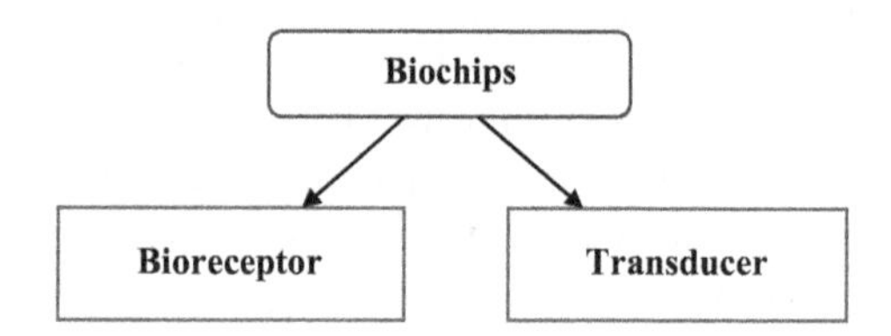

FIGURE 4.2 Schematic representation of a biochip.

4.5 BIOCHIPS

Biochips can be described as "micro-electronic-inspired devices that are used for biological molecules and species." These chips are used to detect biochemically relevant and interesting cells, a variety of microorganisms, proteins, DNA and associated nucleic acids, and minute molecules. The main purpose of the chip is to perform hundreds of biological experiments in a few seconds, as in DNA sequencing. Biochips have a number of single biosensors which can be tracked independently and are used to test numerous analytes. The transducer tests the analyte's contact with bioreceptors, which transforms the information into an observable result, such as an electrical signal (Figure 4.2).

4.6 DIFFERENT TYPES OF BIORECEPTORS

Bioreceptors play a significant role in the detection of substrates because of their specificity. It is responsible for obligating the analyte of interest to the measurement of biosensors. The various bioreceptors have been used in different ways. In general, bioreceptors are categorized into five major groups: (1) antibodies/antigens, (2) enzymes, (3) nucleic acids/DNA, (4) cell structures/cells, and (5) biomimetic and bacteriophages (Figure 4.3).

4.7 TYPES OF TRANSDUCERS

Biosensors are also classified according to the transduction method. Most forms of transduction fall into one of three major classes. These classes are: optical detection

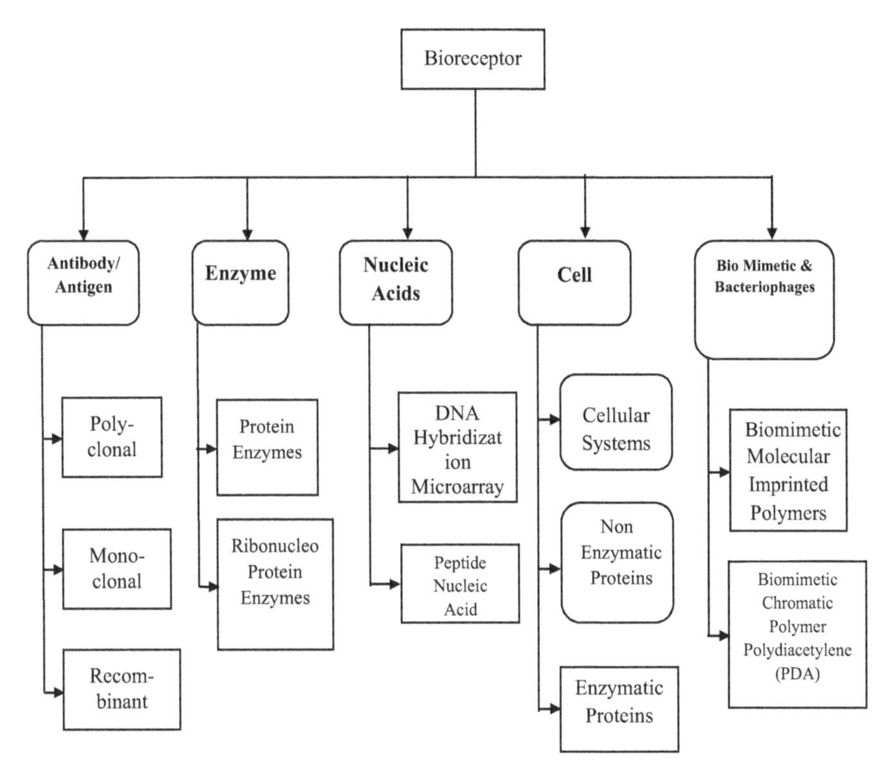

FIGURE 4.3 Schematic representation of a bioreceptor.

methods, electrochemical detection methods, and mass change detection methods [10–12] (Figure 4.4).

Biochips are produced by oligonucleotide or peptide nucleic acid synthesis in situ (on chip) or DNA fragment spotting. Hybridization of nucleic acid-related samples on chips enables the expression of the genomic sequence of mRNA polymorphisms. Biochips enable scientists to scan large numbers of biological analytes rapidly. Affymetrix launched the first commercial chip. Their products for GeneChip include thousands of individual sensors.

The key benefits of biochips are that they are small in size, quick and strong, perform thousands of biological experiments in a few seconds, increase speed of diagnosis, can simultaneously detect multiple viral agents, and are easy to use. In automated microarray manufacturing, new technologies like photolithography, mechanical microspotting and ink jets are used. Bioelectronic microchips have various electronically active microelectrodes connected to the electrodes by molecular wires with specialized DNA capture probes. Several biosensors have also been used in combination with biochips. Instead of performing individual gene assays, these chips are designed to classify multiple genes present in one assay in a sample [11].

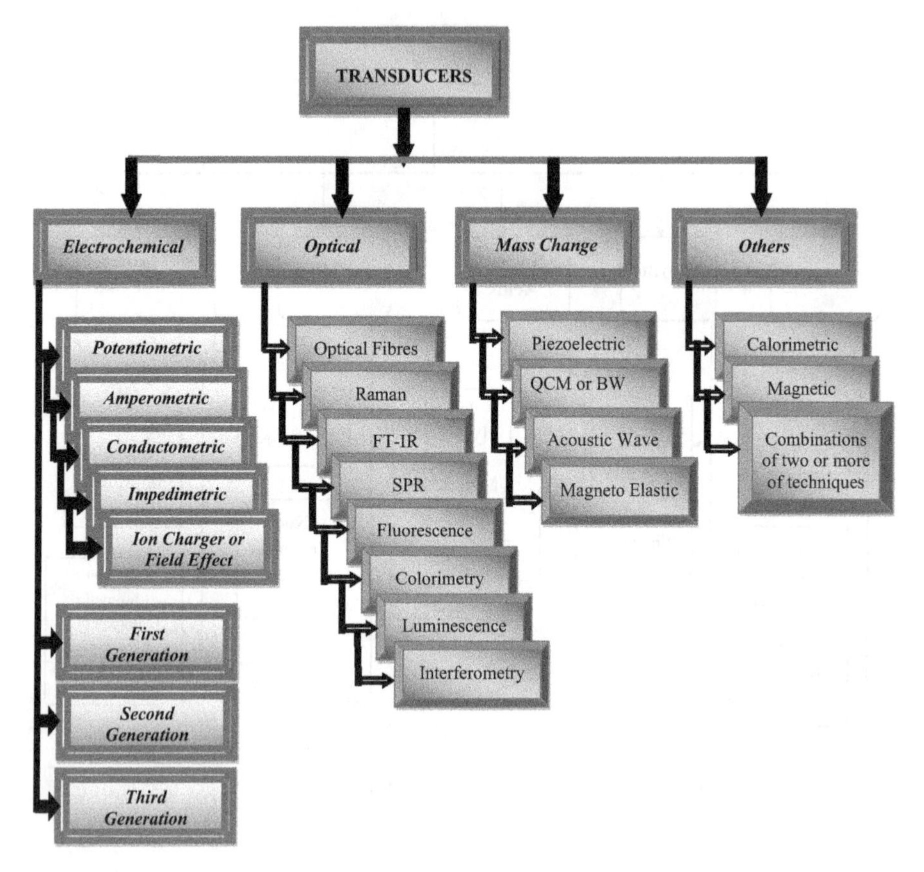

FIGURE 4.4 Classification of transducers.

4.8 WORKING PRINCIPLE OF A BIOCHIP

The following steps are mainly included in the function of a biochip.

1. With radio signals, the operator produces a low-power electromagnetic field.
2. The fixed biochip turns on.
3. The triggered chip transmits the reverse identification code via radio signals to the operator.
4. The reader enhances the received code to convert it into a digital form and then displays it on the LCD.

4.9 COMPONENTS OF A BIOCHIP

A Biochip is comprised of two components, a transponder and a reader.

4.9.1 Transponder

The transponder is a real implant for the biochip. It is composed of a passive transponder that needs a low electrical charge to be activated. The reader "reads" or "scans" the entrenched biochip and receives back information from the biochip. The biochip and reader are linked through decreased-frequency waves. Being passive, the biochip is immobile until the reader triggers it by giving it a decreased-power electrical charge.

The following four sections form the transponder.

4.9.1.1 Antenna coil

- This is generally a simple coil of copper wire around a ferrite or iron core material.
- It is very basic and small, and it is normally used for transmitting and receiving signals from the scanner.

4.9.1.2 Computer microchip

- An exclusive recognition number is processed, ranging from 10 to 15 digits.
- The storage capability of existing microchips is small, and only a single ID number can be stored.
- Before assembly, the specific ID number is "etched" or encoded onto the microchip surface using a laser.
- Once the number is encoded, it cannot be changed.
- The microchip also has requisite electronic circuitry for transmitting the "reader" ID number.

4.9.1.3 Tuning capacitor

- The capacitor stores a tiny electrical charge offered by the reader or scanner, which effectively activates the transponder (less than 1/1000 of a watt).
- This "activation" allows the ID number encoded on the computer chip to be returned by the transponder.
- The capacitor is "tuned" to the same frequency as the reader, since "radio waves" are used to converse between the transponder and reader.

4.9.1.4 Glass capsule

- The glass capsule "houses" the coil, microchip, and capacitor of the antenna.
- It is a tiny capsule about the size of an uncooked grain of rice, varying from 11 mm in length to 2 mm in diameter.
- The capsule is made of soda lime glass that is biocompatible.
- The capsule is given an airtight seal after installation, so no body fluids can enter the electronics inside.
- A material like a polypropylene polymer sheath is applied to one end of the capsule, since the glass is very smooth and susceptible to movement.
- This sheath offers a compatible surface that is bonded to or intertwined with the body tissue fibers, resulting in the biochip being permanently positioned.

4.9.2 Reader

A reader consists of a coil called an exciter, which is used with the aid of radio signals to generate an electromagnetic field. It provides the necessary energy to activate the chip. To receive the transmitted code guided back from the excited implanted chip, a receiving coil is present [12–14].

4.10 BIOCHIP DESIGN

The first step in biochip design is the selection of the biochip substrate. Biochip surfaces are then modified and evaluated to optimize biochip manufacturing processes for diverse test panels. Each biochip has 100–1000 gel drops, each around 100 microns in diameter, with plastic or membrane support. To identify a particular biological agent or biochemical signature, a portion of a DNA strand, protein, peptide, or antibody is inserted into each drop. These exclusive drops are in recognized positions, and the reaction location of a sample can be controlled. The biochip system may be able to recognize infectious disease strains in less than 15 minutes when testing different protein arrays and in less than 2 hours when testing nucleic acid arrays.

4.11 BIOCHIP TYPES

4.11.1 DNA Microarray

A DNA microarray consists of a large number of small spots of DNA that are fixed to a hard surface. It is used to measure the appearance levels for a number of genes. Probes are composed of all DNA marks (the picomoles of a particular gene). In general, the hybridization of the probe target is observed and measured by appreciation of fluorophore-labeled targets in order to establish the relative amount of nucleic acid series in the selected target.

4.11.2 Microfluidic Chip

Microfluidic chips are designed to replace biochemical laboratories. They are used for reactions such as DNA analysis, molecular biology procedures, and many more biochemical reactions. These chips are really complex because they contain thousands of components.

4.11.3 Protein Microarray

These chips are used to monitor protein activities and bonds and to discover their functions on great scale. Their key benefit is that they can be used in tandem to monitor a huge number of proteins. This chip possesses a supporting surface such as the microtiter plate or bead, the glass slip, and the nitrocellulose membrane. These are mechanized, quick, inexpensive, and extremely sensitive, consuming less sample quantities.

4.12 BIOCHIP APPLICATIONS

Biochip technologies are applied in numerous areas, including proteomic, glycomic, and genomic science as well as toxicology and pharmacology. Biochips play an important role in molecular diagnostics, and its use is expected to accelerate the advancement of personalized medicine.

4.12.1 Neural Sensing and Interfacing

In today's world, it is important to interpret brain signals and translate these into control and communication signals for efficient communication with the environment. Such a system is called a brain–computer interface (BCI). Dr. Grey Walter described the first BCI in 1964. Peripheral nerves and muscles are needed for natural modes of communication or control. The individual activates an intricate process based on a specific purpose, in which definite brain areas are stimulated and signals are transmitted to the corresponding muscles through the peripheral nervous system (PNS), which in turn executes the essential group for the task of communication or control. The movement that results from this process is often called motor output or efferent output. "Efferent" refers to the transfer of impulses from the central nervous system (CNS) to the PNS and to effectors. "Afferent" is signaling in the other direction, from the sensory receptors to the CNS. The motor pathway is fundamental for motion control [12].

A substitute to natural control and communication is provided by BCI (see Figure 4.5). A BCI is a synthetic device that bypasses neuromuscular output channels, the body's normal efferent pathways. A BCI specifically monitors brain activity relevant to the intent of the individual and converts the reported brain activity into consequent control signals. The fact that the measured behavior originates in a straight line from the brain and not from the peripheral systems is the reason why the device is known as a brain-computer interface. This conversion requires signal processing and pattern appreciation, which is performed by a computer.

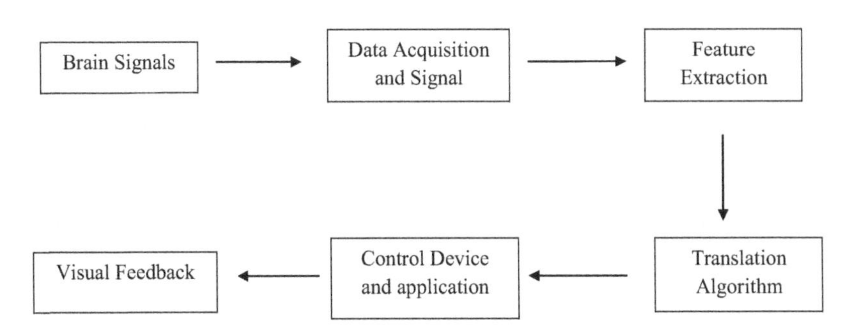

FIGURE 4.5 Schematic representation of a BCI.

4.12.2 Components of a BCI System

A BCI system normally has three major components:

- **a sensor**, which records neural activity or its proxy system;
- **a decoder**, which changes neural activity into command signals; and
- **an effecter**, which is like a computer cursor or robotic arm.

4.12.3 Different Types of BCI

Figure 4.6 shows the different types of BCI and their applications.

4.12.3.1 Invasive BCIs

Invasive BCIs are directly inserted by neurosurgery into the grey matter of the brain so that they can send signals of high quality. But when the body responds to a distant object in the brain, these BCIs are vulnerable to scar tissue building up, causing the signal to become weaker or even be lost.

4.12.3.2 Partially Invasive BCIs

Within the skull, but outside the grey matter, there is another form of brain signal reading phase. Electrocorticography is an instance of partially invasive BCI (electrocorticography). An electrode grid is inserted by surgical incision. It tracks the behavior of the brain within the skull, but from the exterior of the membranes that cover the brain.

4.12.3.3 Non-Invasive BCIs

Electrodes, each of which is connected to an individual wire, are used by several systems. Magnetoencephalography senses the small magnetic fields produced within the brain as individual neurons "fire." It can recognize a millimeter of the active region and can track the flow of brain activity as it travels from region to region within the brain. The variations in hemoglobin's magnetic properties are exploited by functional magnetic resonance imaging (fMRI), as it holds oxygen.

4.12.4 BCI Performance

The efficiency of a BCI can be calculated in different ways. Classification efficiency is a simple metric. It is the dynamic ratio of the number of trials properly classified and the total number of trials. It is also easy to quantify the error rate because it is just the combination of wrongly categorized trials to the total number of trials. While it is easy to calculate classification efficiency or error rates, application-dependent metrics are much more important.

4.12.5 Applications of a BCI

The primary applications of a BCI include:

- providing connectivity, environmental control, and movement rehabilitation for disabled persons;
- providing increased control of equipment for people with disabilities, such as wheelchairs, cars, or assistance robots;
- providing an external control channel for video games;
- tracking treatment for long-distance drivers or aircraft pilots and giving warnings and alerts to aircraft pilots; and
- building intelligent devices for relaxation.

4.13 GENE CHIP ENGINEERING

Depending on the sort of molecule, the biochip is created in two formats (Figure 4.6). cDNA arrays are often referred to as biochips with PCR items of 200 bp to 2 kb scale immobilized by covalently cross-linking to the outside of the array. Alternatively, oligonucleotide probes can either be synthesized in situ on the array, or pre-synthesized oligos can repair covalent terminal contacts. Gene chip engineering involves many separate components such as processing, sample preparation and target sequence hybridization, hybridization effects detection, oligonucleotide probe configuration, and image hybridization analysis, as seen in Figure 4.7. Next, many related gene sequences are chosen from the DNA database for a specific reason. By deciding each probe's sequence and length and its exact location on the chip, a set of special oligonucleotide probes will be built, based on the chosen sequences.

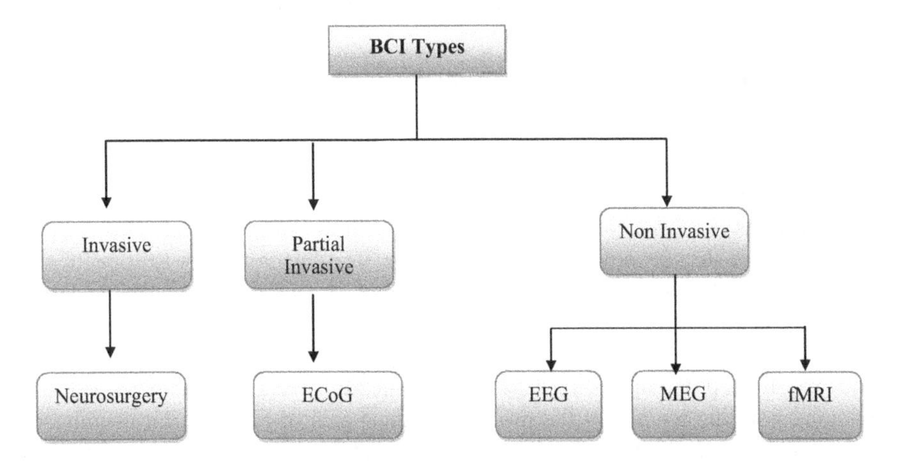

FIGURE 4.6 Types of BCI and their applications.

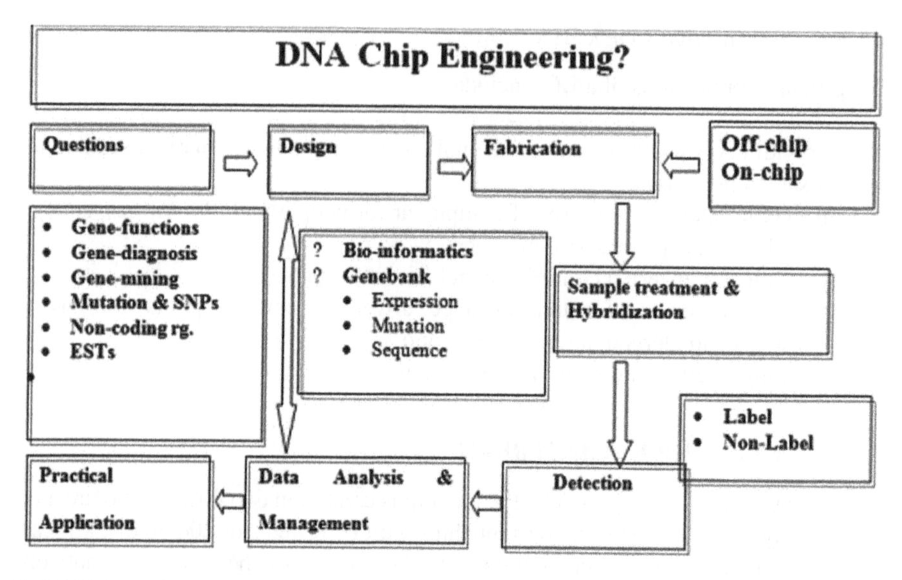

FIGURE 4.7 Several important aspects of gene chip technology.

Using oligonucleotide probes instead of PCR goods has many advantages. First, they are usually of similar length and can be designed to have similar hybridization properties. Second, they may be engineered to hybridize toward the same gene region; PCR also enables cross-hybridization between homologous genes that confuse the gene expression profiles of the same gene family. Third, oligonucleotide arrays remove the need for tedious PCR amplification of probe molecules and decrease the probability of error due to clone handling and contamination when moved.

4.14 BIOINFORMATICS AND GENE CHIP DESIGN

Gene chips may be an accurate and efficient data mining tool. However, the genetic modification and hybridization data shown in Figure 4.8 present several bioinformatics problems. Probe design will influence the quality and precision of gene chips. A series of masks must be designed to correct the positions and orders of each synthetic DNA step for producing a given high-density gene chip. A new mathematical model was developed to avoid the mismatch of target sequences hybridized with probes on the same chip due to the different hybridization denature temperature (Tm) of the probes. An aggregate of optimized probes will be produced for a given target sequence. Each probe in this aggregate is designed to have a consistent Tm value by changing both the lengths of the probes and their area of coverage on neighboring probes. To optimize both oligonucleotide probes and built-in applications capable of obtaining the probes, one method is to modify the overlapping range and duration of the probe sequences with unique sequences to their target and the difference in critical hybridization temperature (TM value) between probes below 0.5 for 256×256 arrays. Optimize the number of candidate samples using dynamic programming arithmetic, and pick the final combination for the gene chip sample sequences. Both chip

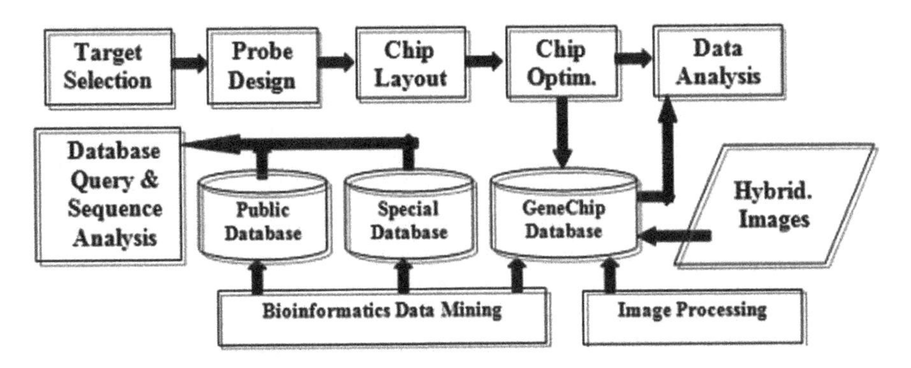

FIGURE 4.8 Bioinformatics problems related to gene chips.

probes have a near-hybridization state with target sequences. Scientists developed various software for high-density gene chip design based on this design process.

4.15 NEURAL CHIPS

In the field of VLSI neural networks, a comprehensive study of application priorities and specifications was carried out to identify and determine complex modular structures for each application, aimed at solving problems of industrial importance, involving real data and operating conditions. The study led to more sophisticated implementation within Elsag Bailey and related companies, particularly for advanced applications such as cursive handwriting, 3D vision, robotic navigation, fault detection, and process control. In some cases, detailed real data tests revealed favorable results relative to conventional state-of-the-art technologies, such as character identification, graphical pattern recognition, image segmentation and classification, robotic navigation, signal processing, fault diagnosis, and factory state estimation. Comprehensive database studies observed a substantial decrease in both the rate of error (without rejection) and the rate of rejection (with equivalent error) relative to the best conventional classifier, particularly for digits and alphabets as opposed to industrial classifiers. In the latter case, the exclusion rate decreased by around 5% compared to commercial classifiers, putting it first in the NIST test results worldwide.

Four classifiers based on simple features work in parallel to create a hybrid hand-printed character recognition system, and their cooperation is used for quality enhancement. The four classifiers are based on two different normalization sequences, two different attribute extraction processes, and two different classification techniques. Three different approaches are used to blend classifier results. The first is based on instructions, the second uses a weighted sum mix based on a perception layer, while the third uses multilayer perception that collects the total data found in the output of the classifiers. Similarly, a 7% increase in number plate position accuracy and an increase of around 3% in character detection accuracy was observed in the graphical pattern recognition program. Data dimensionality reduction in both computation time and accuracy enhancement was accomplished with respect to image processing and recognition, along with a 2–3% improvement in classification performance compared to standard algorithms obtained on the same data. Improvements were also found in radar, voice, biomedical, and sonar signal processing [13].

In the artificial vision domain, where very authentic data can be created, and in robotic navigation, new robust data representation methods have been developed that show remarkable potential for obstacle avoidance and trajectory learning. In the field of plant operation, neural modeling was used to develop a fault diagnostic methodology that could theoretically detect certain previously unknown faults (which is difficult for conventional systems), and a forecast of in-state process control was accomplished with accuracy ten times better than anticipated. These algorithms and chips are being investigated for integration into sensors and biosensors. Related approaches to miniaturization are currently being investigated to study gallium arsenide and other inorganic compounds and to test the physical limitations of atomic-scale incorporation [14].

4.16 SUMMARY

Overall, biochip science and marketing has had an outstanding start with clear momentum, primarily due to the scientific community's hard and creative work and the government funding that has been provided. The future wide-ranging uses and economic advantages of biochip devices are the primary driving force behind this accelerated progress. India's biochip industry should watch closely what's happening in the west. Rapid development has come about from cooperation with experts and biochip firms in other countries. Biochips emerged as a novel forum for biomolecule analysis in the 1980s. There have been innovations in a variety of areas, including life sciences, information processing, microelectronics, and micromechanics.

Biochips are considered valuable potential tools in modern life science testing, medical diagnosis, drug discovery, food safety control, and agriculture, as they perform well and are miniaturized, automatic, and cost-effective. Biochips are considered able to significantly improve the speed and extent of the analytical process and have enormous economic value. Thus in recent years, several governments and manufacturing companies worldwide have invested extensively in this field. However, biochip technology is still in its early development. It's an ever-changing environment, and the industry lacks proprietary core technologies in related fields.

More and more talented scientists, management experts, business people and entrepreneurs will join hands in this extraordinary trend, and India's biochip industry will benefit from a healthy political/social climate, steady economic growth, and high-tech industry production policies. Biochip research and development in India's future is bright. There's furious rivalry here. Whoever regulates the crucial infrastructure will win [13–15].

REFERENCES

1. Verpoorte, E., De Rooij, N.F. 2003. Microfluidics meets MEMS. Proceedings of the IEEE 91(6), 930–953.
2. Convery, N., Gadegaard, N. 2019. 30 years of microfluidics. Micro and Nano Engineering 2, 76–91.

3. Whitesides, G.M. 2006. The origins and the future of microfluidics. Nature 442, 368–373.
4. Astolfi, M., Péant, B., Lateef, M., Rousset, N., Kendall-Dupont, J., Carmona, E., Monet, F., Saad, F., Provencher, D., Mes-Masson, A.-M. 2016. Micro-dissected tumor tissues on chip: An ex vivo method for drug testing and personalized therapy. Lab on a Chip 16(2), 312–325.
5. Bhatia, S.N., Ingber, D.E. 2014. Microfluidic organs-on-chips. Nature Biotechnology 32, 760–772.
6. Huh, D., Hamilton, G.A., Ingber, D.E. 2011. From 3D cell culture to organs-on-chips. Trends in Cell Biology 21(12),745–754.
7. Sun, W., Luo, Z., Lee, J., Kim, H.J., Lee, K., Tebon, P., Feng, Y., Dokmeci, M.R., Sengupta, S., Khademhosseini, A. 2019. Organ-on-a-chip for cancer and immune organs modeling. Advanced Healthcare Materials 8(4), 1801363.
8. Ahadian, S., Civitarese, R., Bannerman, D., Mohammadi, M.H., Lu, R., Wang, E., Davenport-Huyer, L., Lai, B., Zhang, B., Zhao, Y. 2018. Organ-on-a-chip platforms: A convergence of advanced materials, cells, and microscale technologies. Advanced Healthcare Materials 7(2), 1700506.
9. Bhise, N.S., Ribas, J., Manoharan, V., Zhang, Y.S., Polini, A., Massa, S., Dokmeci, M.R., Khademhosseini, A. 2014. Organ-on-a-chip platforms for studying drug delivery systems. Journal of Controlled Release 190, 82–93.
10. Polini, A., Prodanov, L., Bhise, N.S., Manoharan, V., Dokmeci, M.R., Khademhosseini, A. 2014. Organs-on-a-chip: A new tool for drug discovery. Expert Opinion on Drug Discovery 9(4), 335–352.
11. Huh, D., Matthews, B.D., Mammoto, A., Montoya-Zavala, M., Hsin, H.Y., Ingber, D.E. 2010. Reconstituting organ-level lung functions on a chip. Science 328(5986), 1662–1668.
12. Zhang, Y.S., Zhang, Y.-N., Zhang, W. 2017. Cancer-on-a-chip systems at the frontier of nanomedicine. Drug Discovery Today 22(9), 1392–1399.
13. Huh, D., Leslie, D.C., Matthews, B.D., Fraser, J.P., Jurek, S., Hamilton, G.A., Thorneloe, K.S., McAlexander, M.A., Ingber, D.E. 2012. A human disease model of drug toxicity–induced pulmonary edema in a lung-on-a-chip microdevice. Science Translational Medicine 4(159), 159ra147.
14. Liu, W., Sun, M., Lu, B., Yan, M., Han, K., Wang, J. 2019. A microfluidic platform for multi-size 3D tumor culture, monitoring and drug resistance testing. Sensors and Actuators B: Chemical 292, 111–120.
15. Huh, D., Kim, H.J., Fraser, J.P., Shea, D.E., Khan, M., Bahinski, A., Hamilton, G.A., Ingber, D.E. 2013. Microfabrication of human organs-on-chips. Nature Protocols 8(11), 2135–2157.

5 Fabrication of Stretchable Composite Thin Film for Superconductor Applications

Mohammad Harun-Ur-Rashid, Tahmina Foyez, and Abu Bin Imran

CONTENTS

5.1 INTRODUCTION

From the start, the contemporary electronics industry has used metallic oxide-based semiconductor systems fabricated with a combination of metals and semiconductors having outstanding electrical properties and high moduli. In this century, with the growth of advanced technology, the potential applications of flexible electronic devices have created new challenges for conventional electronics and the traditional materials that are used in such devices. Over the last two decades, we the users have witnessed the expeditious rise of small, portable, wearable, stretchable, and soft electronics exhibiting fascinating properties and unique functionalities and offering the best alternatives to long-established inflexible, rigid, brittle, and clumsy electronics. Scientists and technologists from multiple disciplines are accumulating their knowledge and efforts to develop flexible and stretchable electronic devices such as superconductors, circuits, batteries, sensors, solar cells, supercapacitors, and transistors for human- and environment-friendly next-generation electronic applications [1–5].

The most prominent and noteworthy feature of stretchable electronics is the ability to go under mechanical distortions during their operation. Such electronics are used in bioelectronics applications, such as personalized healthcare systems [6], wearable smart displays [7], and implantable prosthetic devices [8]. Superconductors, which have special quantum effects, can transfer electrical energy without any loss and are the essential components for devices such as superconducting nanowire single photon detectors, superconducting quantum interference, and superconducting chips. Basically, most superconducting devices are constructed on rigid substrates, which hampers potential applications where flexibility is mandatory. Naturally occurring stretchable superconducting materials are being used to synthesize supple superconducting devices. However, it has been difficult to improve the devices' dependability and efficiency because of the complicated synthesis process. In such a situation, 2D superconductors based on stretchable composite thin film with mechanical deformability can create a super platform for producing pliable superconducting devices. Superconductive performance, along with stretchability, can be achieved by fabricating superconducting composite thin film capable of exhibiting essential mechanical bendability. This chapter aims to describe different types of artificial stretchable composite with 3D and 2D structures, hybrid stretchable electronics composed of rigid components, and suitable construction materials, structural design, and implementable techniques to fabricate functional stretchable composite thin films. The challenges and future prospects of stretchable composite thin film in superconductive applications are summarized at the end of the chapter.

5.2 NATURE-INSPIRED ARTIFICIAL STRETCHABLE COMPOSITE

Stretchable electronic devices are constructed with outstanding mechanical properties and excellent smart functionalities such as stimuli responsiveness, efficient human–machine interaction, and quick signal processing ability [9, 10]. Bendable displays, stretchable circuits, electronic paper and skin, and wearable power supplies have huge potential for soft robotics, wearable machine interfaces, and medical applications. However, most of these electronics have excellent electrical features but poor stability under complex environmental stress, making them less promising for the fabrication of stretchable electronics. To overcome that limitation, lots of research has already been done to design stretchable and tough electronics which can withstand multidimensional deformations and high environmental stress. Throughout the long history of evolution, the natural components have developed different structures to achieve optimum adaptability in adverse environments. Eventually, the selections of nature allowed researchers to find effective solutions to the drawbacks of stretchable electrical materials [11]. Many scientific publications describe the structural materials which have been patterned by innovations from natural creatures and their activities. For example, by imitating the hierarchical and laminated substructures found in bone, metastability-assisted multiphase steels have been fabricated to attain better fatigue resistance by lowering the chance of crack formation [12]. Spider webs demonstrate excellent mechanical stability through the arrangement of radial and spiral threads having different mechanical properties [13]. Tseng's group created artificial web structures with high mechanical strength using a hierarchically nanostructured assembly [14]. Guo et al. reported a self-charging power unit for harnessing the energy generated by body motion under complex multidirectional deformations like bending, twisting, and stretching [15]. They used gold-graphite nanocomposite and claimed that the self-charging power unit constructed with a kirigami-based supercapacitor (Figure 5.1) harnessed energy from hand movements and supplied the energy to electrical devices.

5.3 ARTIFICIAL STRETCHABLE COMPOSITE IMITATING NATURAL FUNCTIONS

Natural creatures are stable under complex environmental strain and can execute smart activities like reversible locking, mechanical sensing, and dry adhesion. For example, animals and insects' sensory hairs are highly responsive to external stimuli, which motivated researchers to create synthetic electronic whiskers for recognizing and transforming mechanical forces like airflow or other physical contacts [16]. Liu's group designed nanopile interlocking structure (Figure 5.2) inspired by plant root systems, to improve the adhesion between elastic substrates and rigid metal electrodes [17]. They showed that the stretchable electrode might perceive the body-surface biosignals in electromyography, and such a flexible device could be employed for monitoring the distortion prompted by the bending of the elbow. The reported experimental data strongly suggests that their method could be used to fabricate flexible electronics with high adhesion properties.

5.4 STRETCHABLE ELECTRONICS WITH 3D AND 2D STRUCTURES

Stretchable devices must be functional in situations where physical deformations occur due to complex strains. In particular, wearable devices are required to continue to function while struggling with the stretching generated by skin. To match the flexibility of electronic devices with the stretchability of human skin or other surfaces, scientists have designed materials for dissipating strain by altering structural orientation [18–20]. Materials with 3D and 2D structures have been developed to accommodate applied stress through shape changes due to their high moduli and tolerable flexibility. Among the structural variations, buckle-type thin film, also known as wrinkled or wavy design, is the most commonly used structure capable of displaying stretchability. One of the stretchable buckling thin film preparations involves accumulating a steady film on a previously stretched substrate to provide compressive stress along with the wrinkled structure to the film [21]. Other than the normal wrinkled structure, hierarchically buckled materials have also found popularity [22]. Besides different types of buckling structures, structural designs based on origami and kirigami have also been developed to achieve required deformability like bending, twisting, folding, and unfolding at system level [23, 24].

Smart stretchable thin film with patterned structures in 2D configuration is used for fabricating bendable electrical circuits. 2D configured electrodes can be easily fabricated by printing, patterning, and lithography. Thin film composed of metal nanoparticles and elastic substrate composite with 2D structure can be used as

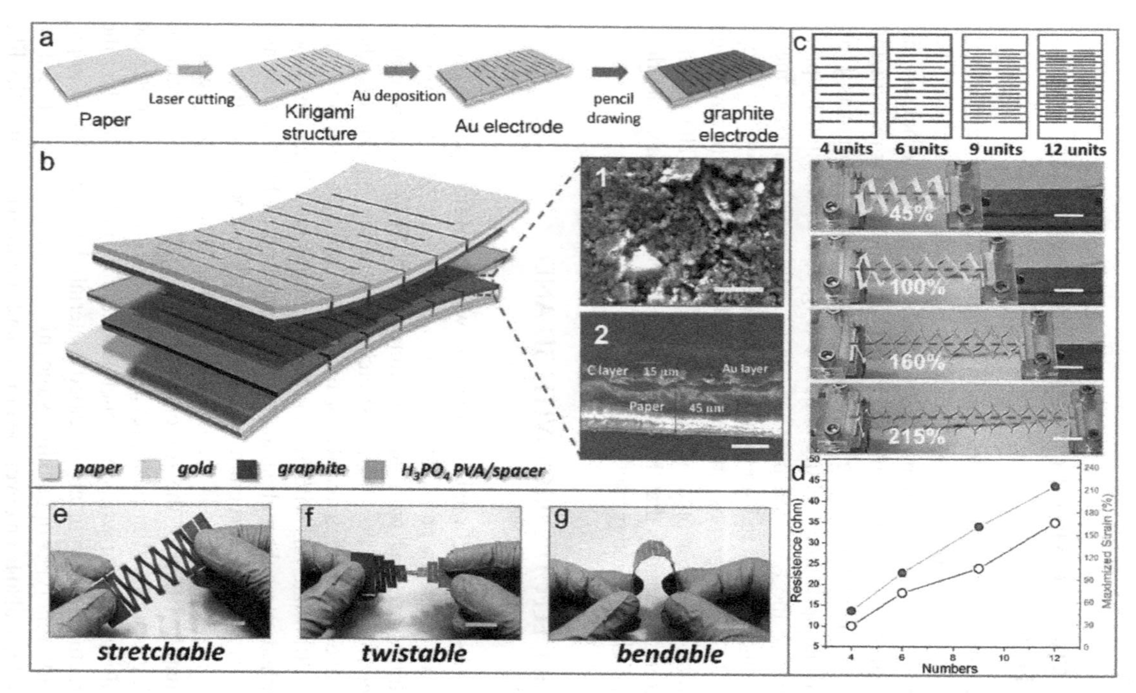

FIGURE 5.1 Schematic illustration and mechanical behaviors of a kirigami-based supercapacitor (a) The electrode preparation process. (b) Schematic framework for the supercapacitor; the scanning electron microscopy (SEM) image of the kirigami-based electrode's (1) cross-section (scale bar 30 μm) and (2) top (scale bar 10 μm) are seen as insets. (c) Schematic diagram of electrodes with various numbers of geometric units (4–12); the associated strain properties of these electrodes are shown in the photographs (scale bar 1 cm). (d) The dependency of resistance and maximum strain with electrodes having different numbers of geometric units. (e–g) Photographs of the supercapacitor under various materials deformations: stretching, twisting, and bending (scale bar 1 cm). The figure has been reproduced with permission from [15] (Copyright 2016, American Chemical Society).

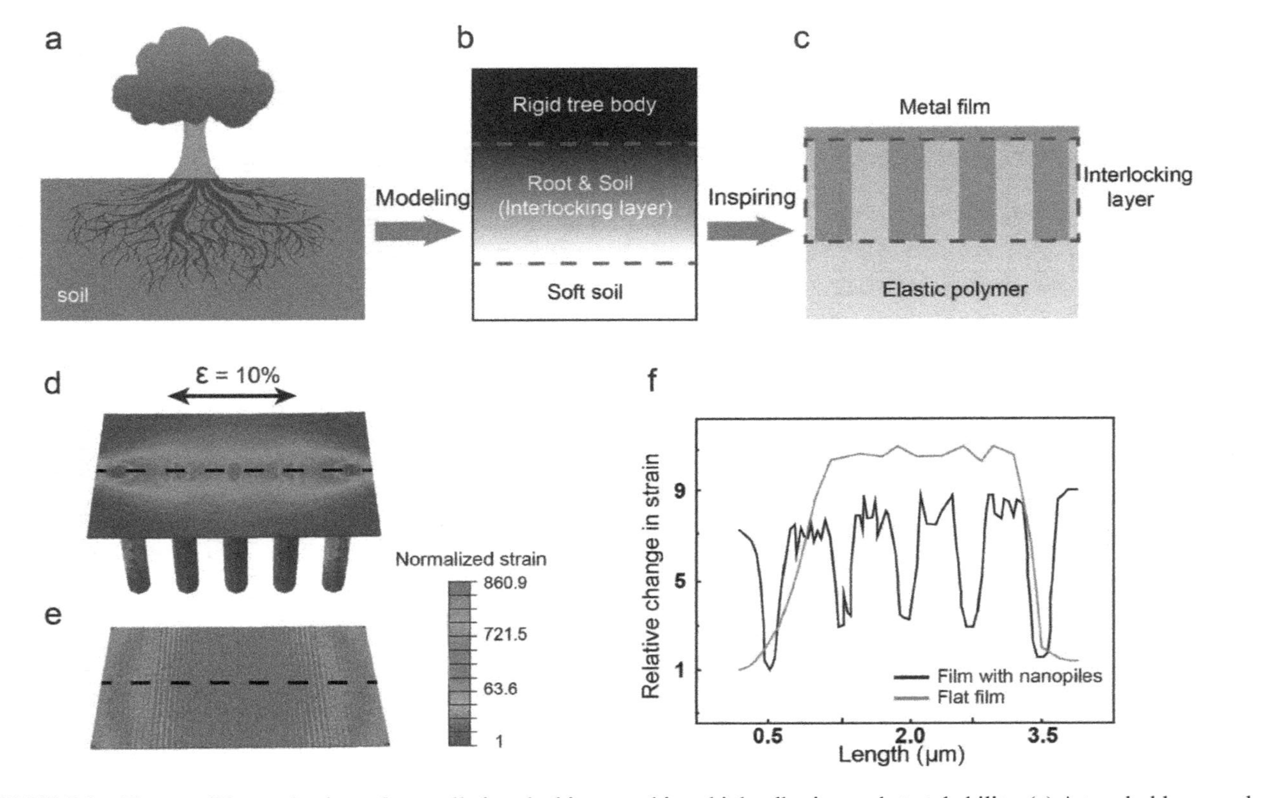

FIGURE 5.2 The possible mechanism of nanopile interlocking to achieve high adhesion and stretchability. (a) A tree holds securely on the ground by extending the fractal roots into the moist soil. (b) A high-adhesion model. (c) Nanopiles are processed under the metal film to form the interlocking layer. (d, e) Finite element modeling simulations in the film with and without nanopiles under the 10% tensile strain. (f) Strain distribution along the dashed black lines in (d) and (e). The figure has been reproduced with permission from [17] (Copyright 2017, John Wiley and Sons).

stretchable electrodes by controlling the size and shape of metal particles and introducing nanocracks and microcracks [25].

5.5 HYBRID STRETCHABLE ELECTRONICS

Scientists have already made substantial efforts in the fabrication of stretchable electronic devices. However, numerous operating devices are composed of fragile functional components and sophisticated architectures sensitive to any sort of deformability. In such a situation, both the efficiency and bendability of devices are retained by hybridizing flexible and inflexible materials. Rigid islands connected by pliable interconnections, one of the mentionable examples of hybrid stretchable architectures, are fabricated by applying patterns [26], buckling [27], and fractal [28] approaches to materialize stretchable circuits with super electrical conductivity.

5.6 MATERIALS FOR STRETCHABLE COMPOSITE THIN FILM

Stretchable conductive polymers are extensively studied because of their electrical conductivity, versatile tenability, and easy and scalable processing for many special electrical and optoelectronic applications [29]. At present, there are lots of accessible conductive polymers, including polyaniline, polyacetylene, polythiophene, polypyrrole, and their derivatives. Poly(3,4-ethylenedioxythiophene) polystyrene sulfonate is the most popular and promising conductive polymer because of its outstanding electrical conductivity, chemical suitability, availability, transparency, and biocompatibility. The conductivity of this polymer could be improved by doping. Nanomaterials based on rubber composites are processed by percolating conductive fillers such as nanoparticles, nanowires, nanosheets, and nanotubes. Polydimethylsiloxane (PDMS) is extensively utilized as the substrate for fabricating stretchable electronics due to its chemical and physical stability, better thermal properties, biocompatibility, and structural modification capability [30]. Ion gel composed of porous polymeric matrix is another promising material for fabricating stretchable superconductive thin film because of its higher ionic conductivity, physical and chemical stability, suitable specific capacitance, and thermal stability [31].

High electrical conductivity and charge–transport capability along with mechanical deformability are the standard criteria for electronic filler materials to fabricate intrinsically stretchable composite thin films. The porous nanostructures of supporting materials facilitate the free movement of electrical fillers through percolation even under mechanical deformation. Metal-based nanofillers are promising percolating materials due to their high conductivity and malleability. Silver nanomaterials, especially Ag nanowires, are commonly used as nanofillers because of their easy synthetic route, higher conductivity, and cost-effective manufacturing techniques.

The electrical conductivity of stretchable thin films depends on the type of nanofiller; however, the films' mechanical characteristics are greatly influenced by elastomer, which provides support during mechanical deformation under stress. Physically crosslinked elastomers such as silicone rubbers (PDMS and Ecoflex), chemically crosslinked elastomers such as styrene elastomers (polystyrene–butadiene–styrene or polystyrene-co-ethylene butylene-co-styrene), and physicochemically crosslinked elastomers such as polyurethane-based elastomers are widely used in fabricating stretchable composite thin films.

5.7 FABRICATION OF FUNCTIONAL STRETCHABLE COMPOSITE THIN FILMS

Feng's group has reported an easy method for fabricating graphene oxide(GO)/chitosan composite film using spin coating [32]. They claimed that composite thin film could be used as a dielectric spacer for superconductor applications and gate dielectrics for electric double layer transistors (Figure 9.3). A GO solution was prepared by dispersing GO sheets in water at a concentration of 0.5 mg/mL. Then a chitosan–acetic acid solution (2.0 wt%) was prepared. The GO and chitosan solutions (1:1 volume ratio) were then mixed together by magnetic stirring. A spin-coater at a speed of 500 rpm on a substrate was used to form GO/chitosan composite stretchable thin film from the solution.

Thin film based on stretchable superconducting yarn, composed of niobium nitride (NbN) nanowire thin fibers on carbon nanotube (CNT) sheets, was fabricated using a sputter deposition technique followed by spinning and twisting [33]. The experimental evidence showed that the NbN–CNT composite yarns could keep their superconductive nature even under extreme mechanical deformations due to excellent flexibility attributed by the nanowire-based porous structure.

Chi et al. fabricated superconducting nanowire single-photon detectors based on the fractal design of the nanowires [34]. They designed niobium-titanium nitride (NbTiN) thin films by patterning Peano curves with lines of 100 nm and merged the nanowires with optical microcavities to increase optical absorption. They deposited 9 nm thick NbTiN film for 82 seconds using reactive magnetron co-sputtering from two highly pure targets of Nb and Ti in a nitrogen–argon gas mixture at room temperature. Using electron-beam lithography, thc nanowires were designed to withstand a high voltage of 100 kV. A 150 nm-thick layer of polymethylmethacrylate was used as a resist. The technique of fractal structures has been applied in fabricating mechanically stretchable interconnects to conduct direct current [35] and antennae for receiving and transmitting microwave signals [36].

Han and colleagues fabricated stretchable composite thin film superconducting device on wafer-scale 2D lead (Pb) nano-islands and single-crystalline graphene on a flexible parylene C substrate [37]. They devised a universal method to fabricate wafer-scale 2D stretchable superconducting device arrays by setting down distinct Pb nano-islands on highly flexible graphene/parylene composite film. The graphene-based single-crystalline layer was formulated on the pure Ge(110) substrate using the chemical vapor deposition technique, and a 1 mm-thick layer of parylene C was deposited by vaporization. The flexible graphene/parylene thin film was then separated by etching, and the Ti–Au (10 nm and 100 nm, respectively) electrodes were deposited on the graphene/parylene film using a Hall bar. The investigators claimed that the fabricated devices were capable of exhibiting superconductivity even under physical deformations. Due to the dielectric characteristic of parylene film, the superconductivity of the stretchable device can be precisely and finely controlled by

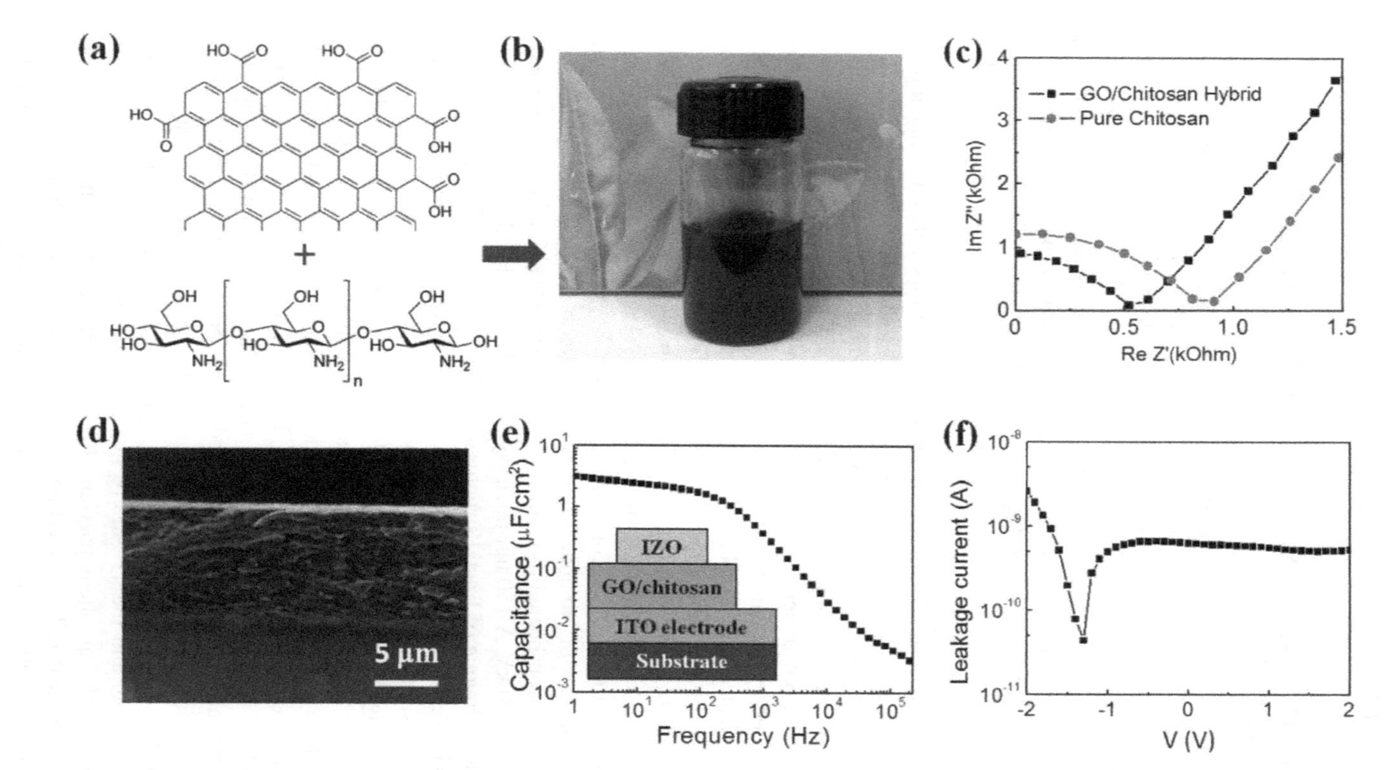

FIGURE 5.3 From GO sheets and chitosan to the composite GO/chitosan and the characterization of the composite electrolyte film. (a) Chemical structures of GO sheet and chitosan. (b) GO sheets and chitosan in water. (c) Traditional Nyquist plot of GO/chitosan composite film (black dots). (d) Cross-sectional SEM image of the GO/chitosan composite electrolyte film (scale bar 5 µm). (e) The specific capacitance vs. frequency plot of GO/chitosan composite electrolyte film. (f) Leakage current of the GO/chitosan composite electrolyte film. The figure has been reproduced with permission from [32] (Copyright 2016, Springer Nature).

monitoring the gate voltage. The combination of separated superconducting islands and stretchable graphene/parylene offers a feasible way to construct state-of-the-art superconducting devices.

A schematic illustration of the fabrication process of the wafer-scale flexible 2D superconductive device arrays is shown in Figure 5.4.

Jia et al. proposed a controllable microfluidic fabrication technique for stretchable and wearable carbon black/graphene composite fibers for superconductor applications [38]. The fabrication process is demonstrated in Figure 5.5a. The superconductive carbon black/graphene composite fibers (SCB/GFs) possessed excellent mechanical flexibility and physical deformability, which can be used to formulate a stretchable fiber spring (Figure 5.5b). The composite fibers were able to produce flexible supercapacitors with high specific capacitance. The fabricated stretchable supercapacitors with full charge were employed to power an electronic timer and light emitting diode lamps. The research group suggested that the stretchable supercapacitors would find potential applications in stretchable and wearable superconducting devices or in flexible electrodes for constructing energy storage and conversion devices.

5.8 APPLICATIONS

Stretchable composite thin films may find applications in fabricating intrinsically stretchable sensors, actuators, transistors, light emitting devices, and stretchable bioelectronic systems. Sensors, especially stretchable mechanical sensors, are essential units of flexible electrical appliances for imitating the mechanoreceptors of human skin [39] or monitoring the movement of the human body [40]. Actuators, particularly associated with the fabrication of sensors, are the most widely studied actuating devices. Hong's group constructed a transparent and flexible heater utilizing silver nanowires and PDMS composite through heterogeneous rearrangement techniques [41]. The stretchable transistor is one of the fundamental components of stretchable electronic devices. Wang et al. fabricated a stretchable transistor composed of polymeric dielectric and active channel layers for stretchable electrical applications [42]. Liang et al. reported on the fabrication of stretchable polymeric light-emitting devices using a flexible composite of GO, soldered silver nanowires, and UV-treated polyurethane acrylate elastomers [43]. Son and his group designed and constructed a multifaceted stretchable bioelectronic system combined with a wearable strain sensor, flexible ECG sensors, and light-emitting capacitor by using self-healing stretchable polymer composite thin film [44].

5.9 SUMMARY AND PERSPECTIVES

The conventional, and most commonly used, methods for fabricating stretchable thin film for electronic applications are solution-spinning, wet-spinning, and coaxial spinning [45–47]. This literature review suggests that GO nanosheets would be severely restacked during fabrication due to the strong π-interaction, which curbs the unique features of graphene and highly limits their prospective applications [48]. Interestingly, as a prominent and effective technique to fabricate

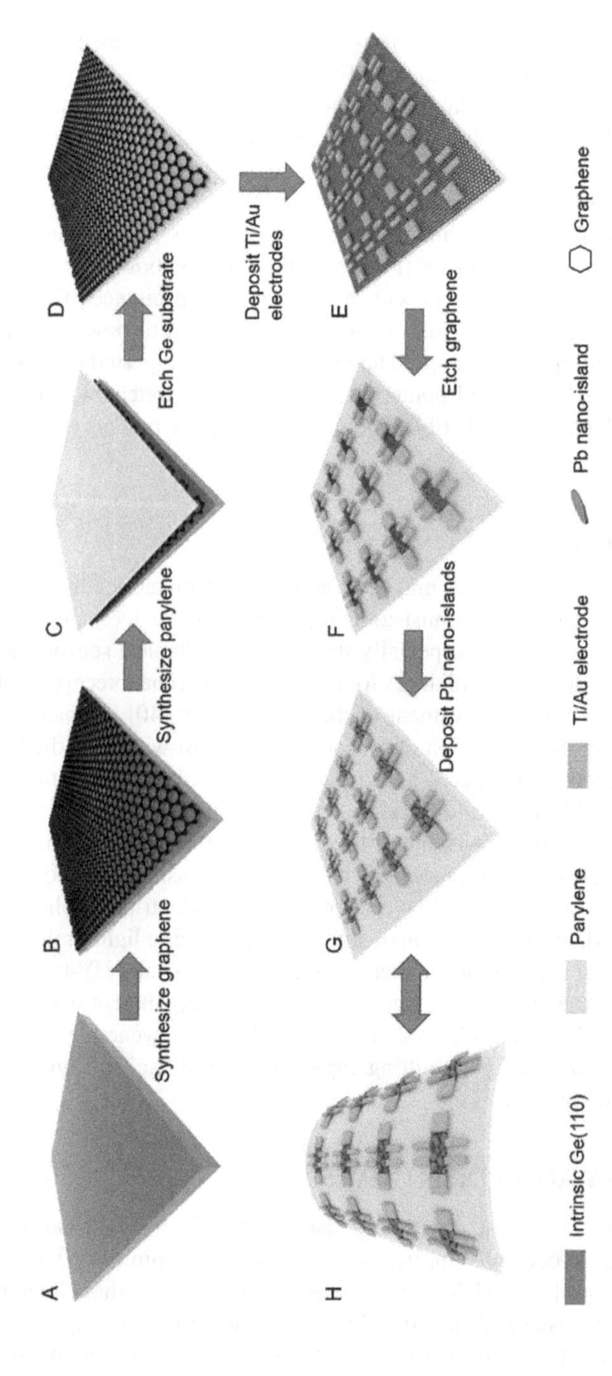

FIGURE 5.4 Schematic illustration of the fabrication process of the wafer-scale flexible 2D superconductive device arrays. The figure has been reproduced with permission from [37] (Copyright 2020, Royal Society of Chemistry).

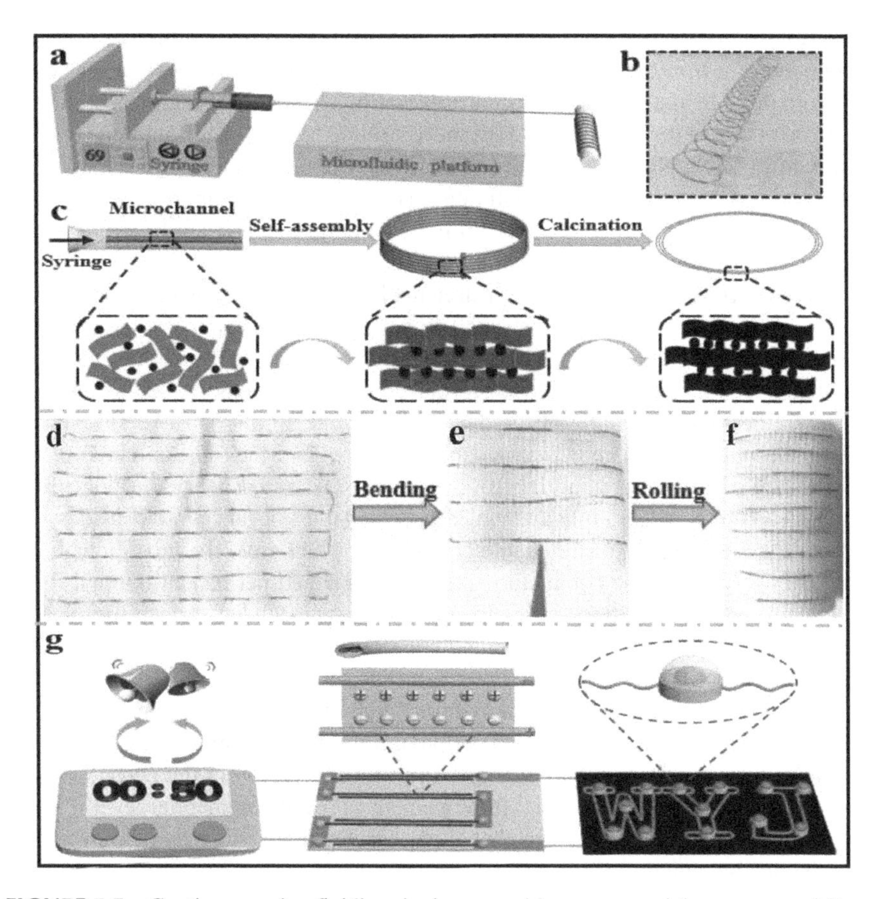

FIGURE 5.5 Continuous microfluidic spinning assembly process and the as-prepared fibers. (a) Schematic representation of the formulation of SCB/GFs from a microfluidic-fabricated technique. (b) The freestanding fiber spring. (c) Schematic representation of the porous structure formed between nanosheets of SCB/GFs. (d–f) Images of cotton fabric woven fibers, cotton fabric bending, and the rolling system. (g) Carbon/graphene supercapacitors used to power various electronic devices. The figure has been reproduced with permission from [38] (Copyright 2020, Elsevier).

stretchable composite thin film, the microfluidic spinning method has become very attractive to researchers [49]. Compared with conventional spinning strategies, the microfluidic approach possesses various exceptional features such as fine tenability, multichannel reaction, and non-hazardous operational conditions [50]. In particular, the ceaseless microfluid can make films or fibers with a fixed structure through self-assembly, which provides desirable characteristics, including superconductivity, greater areal energy density, and high specific capacitance [51]. Desirable functional properties might be incorporated by introducing some active materials like amorphous carbon, CNT, and metallic oxides [52, 53]. Surprisingly, the introduction of carbon black adds the property of superconductive to the fabricated composite [54]. The superconductive nature of composite thin film can be precisely tuned and converted into stretchable, wearable, and flexible material for practical applications in stretchable superconducting electronics. Polyrotaxane-based, extremely stretchable hydrogels and elastomeric films would be excellent candidates when designing thin film for superconductors [55–57]. It is imperative to investigate easy and effective approaches for fabricating stretchable composite thin film for superconductor applications in the above circumstances.

Flexible electronics constructed on stretchable materials provide many advantages because of their extraordinary physical properties and mechanical characteristics. There are many applications of stretchable superconducting thin films; however, producing such devices with high performance is still challenging due to poor carrier mobilities, larger volume of the device, lower integration density, and parasitic effects. Despite huge potential, stretchable thin films and their accompanying technologies are still immature and far from commercialization. The stability of flexible electronic systems is also a fundamental issue which must be taken into consideration. Use of silver nanowires in stretchable conductors, which hold the most potential as candidates for stretchable conductors, face self-generated permanent internal resistance due to continuous stress–strain cycles over their extended service life. Several techniques, like soldering, annealing, and pressing, have already been employed, but these do not solve the problem. Choi et al.'s article showed that antioxidant treatment of the surface of silver nanowires led to improved performance resulting in superconductivity and stability for an extended period of time [58]. The stretchable technology of wireless communication and energy harnessing devices is in the early stages of development and needs substantial improvement to reach commercial level. Further technological advancement of stretchable electrical components may help overcome the existing challenges.

ACKNOWLEDGMENTS

A.B. Imran gratefully acknowledges the Ministry of Education, People's Republic of Bangladesh for the funding for this review.

REFERENCES

1. Yin, L., Lv, J. and Wang, J. 2020. Structural innovations in printed, flexible, and stretchable electronics. *Advanced Materials Technologies* 5(11): 2000694. doi:10.1002/admt.202000694
2. Lu, Y., Xu, K. Zhang, L. Deguchi, M. Shishido, H. Arie, T., Pan R., Hayashi, A., Shen, L., Akita, S. et al. 2020. Multimodal plant healthcare flexible sensor system. *ACS Nano* 14(9): 10966–10975.
3. Matsuhisa, N., Chen. X., Bao. Z. and Someya. T. 2019. Materials and structural designs of stretchable conductors. *Chemical Society Reviews* 48(11): 2946–2966.
4. Kim, D. C., Shim, H. J., Lee, W., Koo, J. H. and Kim, D.-H. 2019. Material-based approaches for the fabrication of stretchable electronics. *Advanced Materials* 32(15): 1902743. doi:10.1002/adma.201902743

5. Amjadi, M., Kyung, K.-U., Park, I. and Sitti, M. 2016. Stretchable, skin-mountable, and wearable strain sensors and their potential applications: A review. *Advanced Functional Materials* 26(11): 1678–1698.
6. Lee, H., Choi, T. K., Lee, Y. B., Cho, H. R., Ghaffari, R., Wang, L., Choi, H. J., Chung, T. D., Lu, N., Hyeon, T. et al. 2016. A graphene-based electrochemical device with thermoresponsive microneedles for diabetes monitoring and therapy. *Nature Nanotech* 11: 566–572.
7. Choi, M. K., Yang, J., Kim, D. C., Dai, Z., Kim, J., Seung, H., Kale, V. S., Sung, S. J., Park, C. R., Lu, N. et al. 2017. Extremely vivid, highly transparent, and ultrathin quantum dot light-emitting diodes. *Advanced Materials* 30(1): 1703279. doi:10.1002/adma.201703279
8. Park, S., Heo, S. W., Lee, W., Inoue, D., Jiang, Z., Yu, K., Jinno, H., Hashizume, D., Sekino, M., Yokota, T. et al. 2018. Self-powered ultra-flexible electronics via nano-grating-patterned organic photovoltaics. *Nature* 561(7724): 516–521.
9. Liu, Y., He, K., Chen, G., Leow, W. R. and Chen, X. 2017. Nature-inspired structural materials for flexible electronic devices. *Chemical Reviews* 117(20): 12893–12941.
10. Bao, Z. and Chen, X. 2016. Flexible and stretchable devices. *Advanced Materials* 28(22): 4177–4179.
11. Park, J., Lee, Y., Ha, M., Cho, S. and Ko, H. 2016. Micro/nanostructured surfaces for self-powered and multifunctional electronic skins. *Journal of Material Chemistry B* 18: 2999–3018.
12. Koyama, M., Zhang, Z., Wang, M., Ponge, D., Raabe, D., Tsuzaki, K., Noguchi, H. and Tasan, C. C. 2017. Bone-like crack resistance in hierarchical metastable nanolaminate steels. *Science* 355(6329): 1055–1057.
13. Lin, L. H., Edmonds, D. T. and Vollrath, F. 1995. structural engineering of an orb-spider's web. *Nature* 373: 146–148.
14. Tseng, P., Napier, B., Zhao, S., Mitropoulos, A. N., Applegate, M. B., Marelli, B., Kaplan, D. L. and Omenetto, F. G. 2017. Directed assembly of bioinspired hierarchical materials with controlled nanofibrillar architectures. *Nature Nanotechnology* 12: 474–480.
15. Guo, H., Yeh, M.-H., Lai, Y.-C., Zi, Y., Wu, C., Wen, Z., Hu, C. and Wang, Z. L. 2016. All-in-one shape-adaptive self-charging power package for wearable electronics. *ACS Nano* 10(11): 10580–10588.
16. Takei, K., Yu, Z., Zheng, M., Ota, H., Takahashi, T. and Javey, A. 2014. Highly sensitive electronic whiskers based on patterned carbon nanotube and silver nanoparticle composite films. *Proceedings of the National Academy of Sciences of the United States of America* 111: 1703–1707.
17. Liu, Z., Wang, X., Qi, D., Xu, C., Yu, J., Liu, Y., Jiang, Y., Liedberg, B. and Chen, X. 2017. High-adhesion stretchable electrodes based on nanopile interlocking. *Advanced Materials* 29(2): 1603382. doi:10.1002/adma.201603382
18. Liu, W., Song, M. S., Kong, B. and Cui, Y. 2017. Flexible and stretchable energy storage: Recent advances and future perspectives. *Advanced Materials* 29(1): 1603436. doi:10.1002/adma.201603436
19. Wang, Y., Li, Z. and Xiao, J. 2016. Stretchable thin film materials: Fabrication, application, and mechanics. *Journal of Electronic Packaging*, 138(2): 020801. https://doi.org/10.1115/1.4032984.
20. Harris, K. D., Elias, A. L. and Chung, H. J. 2016. Flexible electronics under strain: A review of mechanical characterization and durability enhancement strategies. *Journal of Material Science* 51: 2771–2805.
21. Cheng, H., Zhang, Y., Hwang, K.-C., Rogers, J. A. and Huang, Y. 2014. Buckling of a stiff thin film on a pre-strained bi-layer substrate. *International Journal of Solids and Structures* 51(18): 3113–3118.

22. Mu, J., Hou, C., Wang, G., Wang, X., Zhang, Q., Li, Y., Wang, H. and Zhu, M. 2016. An elastic transparent conductor based on hierarchically wrinkled reduced graphene oxide for artificial muscles and sensors. *Advanced Materials* 28(43): 9491–9497.
23. Iwata, Y. and Iwase, E. 2017. Stress-free stretchable electronic device using folding deformation. In *Micro Electro Mechanical Systems (MEMS)*, IEEE 30th International Conference: 231–234.
24. Rafsanjani, A. and Bertoldi, K. 2017. Buckling-induced kirigami. *Physical Review Letter* 118: 084301.
25. Adrega, T. and Lacour, S. P. 2010. Stretchable gold conductors embedded in PDMS and patterned by photolithography: Fabrication and electromechanical characterization. *Journal of Micromechanics and Microengineering* 20(5): 055025. 10.1088/0960-1317/20/5/055025
26. Kim, D. H., Lu, N., Ghaffari, R., Kim, Y. S., Lee, S. P., Xu, L., Wu, J., Kim, R. H., Song, J., Liu, Z. et al. 2011. Materials for multifunctional balloon catheters with capabilities in cardiac electrophysiological mapping and ablation therapy. *Nature Materials* 10: 316–323.
27. Ko, H. C., Stoykovich, M. P., Song, J., Malyarchuk, V., Choi, W. M., Yu, C. J., Geddes, J. B., III, Xiao, J., Wang, S., Huang, Y. et al. 2008. A hemispherical electronic eye camera based on compressible silicon optoelectronics. *Nature* 454: 748–753.
28. Xu, S., Zhang, Y., Cho, J., Lee, J., Huang, X., Jia, L., Fan, J. A., Su, Y., Su, J., Zhang, H. et al. 2013. Stretchable batteries with self-similar serpentine interconnects and integrated wireless recharging systems. *Nature Communications* 4: 1543.
29. Sim, K., Rao, Z., Ershad, F. and Yu, C. 2019. Rubbery electronics fully made of stretchable elastomeric electronic materials. *Advanced Materials* 32(15): 1902417. doi:10.1002/adma.201902417
30. Qi, D., Zhang, K., Tian, G., Jiang, B. and Huang, Y. 2020. Stretchable Electronics Based on PDMS Substrates. *Advanced Materials* 33(6): 2003155. doi:10.1002/adma.202003155
31. Le Bideau, J., Viau, L. and Vioux, A. 2011. Ionogels, ionic liquid based hybrid materials. *Chemical Society Reviews* 40(2): 907–925.
32. Feng, P., Du, P., Wan, C., Shi, Y. and Wan, Q. 2016. Proton conducting graphene oxide/chitosan composite electrolytes as gate dielectrics for new-concept devices. *Scientific Reports* 6(1): 34065. doi:10.1038/srep34065
33. Kim, J.-G., Kang, H., Lee, Y., Park, J., Kim, J., Truong, T. K., Kim, E. S., Yoon, D. H., Lee, Y. H. and Suh, D. 2017. Carbon-nanotube-templated, sputter-deposited, flexible superconducting NbN nanowire yarns. *Advanced Functional Materials* 27(30): 1701108. doi:10.1002/adfm.201701108
34. Chi, X., Zou, K., Gu, C., Zichi, J., Cheng, Y., Hu, N., Lan, X., Chen, S., Lin, Z., Zwiller, V. and Hu, X. 2018. Fractal superconducting nanowire single-photon detectors with reduced polarization sensitivity. *Optics Letters* 43(20): 5017–5020.
35. Fan, J. A., Yeo, W., Su, Y., Hattori, Y., Lee, W., Jung, S. Y., Zhang, Y., Liu, Z., Cheng, H., Falgout, L., Bajema, M., Coleman, T., Gregoire, D., Larsen, R. J., Huang, Y. and Rogers, J. A. 2014. Fractal design concepts for stretchable electronics. *Nature Communications* 5(1): 3266. doi:10.1038/ncomms4266.
36. Werner, D. H., Haupt, R. L. and Werner, P. L. 1999. Fractal antenna engineering: The theory and design of fractal antenna arrays. *IEEE Antennas and Propagation Magazine* 41(5): 37–58.
37. Han, X., Gao, M., Wu, Y., Mu, G., Zhang, M., Mei, Y., Chu, P. K., Xie, X., Hu, T. and Di, Z. 2020. Gate-tunable two-dimensional superconductivity revealed in flexible wafer-scale hybrid structures. *Journal of Materials Chemistry C* 41: 14605–14610.

38. Jia, Y., Ahmed, A., Jiang, X., Zhou, L., Fan, Q. and Shao, J. 2020. Microfluidic fabrication of hierarchically porous superconductive carbon black/graphene hybrid fibers for wearable supercapacitor with high specific capacitance. *Electrochimica Acta*, 354: 136731. doi:10.1016/j.electacta.2020.136731
39. Hong, S., Lee, S. and Kim, D. 2019. Materials and design strategies of stretchable electrodes for electronic skin and its applications. *Proceedings of the IEEE* 107(10): 2185–2197. doi: 10.1109/JPROC.2019.2909666.
40. Jung, S., Hong, S., Kim, J., Lee, S., Hyeon, T., Lee, M. and Kim, D.-H. 2015. Wearable fall detector using integrated sensors and energy devices. *Scientific Reports* 5: 17081. https://doi.org/10.1038/srep17081
41. Hong, S., Lee, H., Lee, J., Kwon, J., Han, S., Suh, Y. D., Cho, H., Shin, J., Yeo, J. and Ko, S. H. 2015. Highly stretchable and transparent metal nanowire heater for wearable electronics applications. *Advanced Materials* 27(32): 4744–4751.
42. Wang, S., Xu, J., Wang, W., Wang, G.-J. N., Rastak, R., Molina-Lopez, F., Chung, J. W., Niu, S., Feig, V. R., Lopez, J. et al. 2018. Skin electronics from scalable fabrication of an intrinsically stretchable transistor array. *Nature* 555: 83–88.
43. Liang, J., Li, L., Tong, K., Ren, Z., Hu, W., Niu, X., Chen, Y. and Pei, Q. 2014. Silver nanowire percolation network soldered with graphene oxide at room temperature and its application for fully stretchable polymer light-emitting diodes. *ACS Nano* 8(2): 1590–1600.
44. Son, D., Kang, J., Vardoulis, O., Kim, Y., Matsuhisa, N., Oh, J. Y., To, J. W., Mun, J., Katsumata, T., Liu, Y. et al. 2018. An integrated self-healable electronic skin system fabricated via dynamic reconstruction of a nanostructured conducting network. *Nature Nanotechnology* 13(11): 1057–1065.
45. Meng, F., Lu, W., Li, Q., Byun, J., Oh, Y. and Chou, T. 2015. Graphene-based fibers: A review. *Advanced Materials* 27(35): 5113–5131.
46. Li, J., Li, J., Li, L., Yu, M., Ma, H. and Zhang, B. 2014. Flexible graphene fibers prepared by chemical reduction-induced self-assembly. *Journal of Material Chemistry A* 2(18): 6359–6362.
47. Sun, J., Li, Y., Peng, Q., Hou, S., Zou, D., Shang, Y., Li, Y., Li, P., Du, Q., Wang, Z., Xia, Y., Xia, L., Li, X. and Cao, A. 2013. Macroscopic, flexible, high-performance graphene ribbons. *ACS Nano* 7(11): 10225–10232.
48. Li, M., Tang, Z., Leng, M. and Xue, J. 2014. Flexible solid-state supercapacitor based on graphene-based hybrid films. *Advanced Functional Materials* 24(47): 7495–7502.
49. Wu, G., Tan, P., Wu, X., Peng, L. and Chen, S. 2017. High-performance wearable micro-supercapacitors based on microfluidic-directed nitrogen-doped graphene fiber electrodes. *Advanced Functional Materials* 27(36): 1702493–1702504.
50. Wu, X., Tan, F., Cheng, H., Hong, R., Wang, F., Wu, G. and Chen, S. 2018. Construction of microfluidic-oriented polyaniline nanorod arrays/graphene composite fibers towards wearable micro-supercapacitors. *Journal of Material Chemistry A* 6: 8940–8946.
51. Qu, G., Cheng, J., Li, X., Yuan, D., Chen, P., Chen, X., Wang, B. and Peng, H. 2016. A fiber supercapacitor with high energy density based on hollow graphene/conducting polymer fiber electrode. *Advanced Materials* 28(19): 3646–3652.
52. Rezaul Karim, M., Harun-Ur-Rashid, M. and Imran, A. 2020. Highly stretchable hydrogel using vinyl modified narrow dispersed silica particles as cross-linker. *Chemistry Select* 5(34), 10556–10561.
53. Rahman, A., Solaiman, F. T., Susan, M. and Imran, A. 2020. self-healable and conductive double-network hydrogels with bioactive properties. *Macromolecular Chemistry and Physics*: 2000207. doi:10.1002/macp.202000207.

54. Marriam, I., Wang, X., Tebyetekerwa, M., Chen, G., Zabihi, F., Pionteck, J., Peng, S., Ramakrishna, S., Yang, S. and Zhu, M. 2018. Bottom-up approach to design wearable and stretchable smart fibers with organic vapor sensing behaviors and energy storage properties. *Journal of Material Chemistry A* 6: 13633–13643.
55. Bin Imran, A., Esaki, K., Gotoh, H., Seki, T., Ito, K., Sakai, Y. and Takeoka, Y. 2014. Extremely stretchable thermosensitive hydrogels by introducing slide-ring polyrotaxane cross-linkers and ionic groups into the polymer network. *Nature Communications* 5: 5124.
56. Gotoh, H., Liu, C., Imran, A., Hara, M, Seki, T., Mayumi, K., Ito, K. and Takeoka, Y. 2018. Optically transparent, high-toughness elastomer using a polyrotaxane cross-linker as a molecular pulley. *Science Advances* 4: eaat7629.
57. Imran, A., Harun-Ur-Rashid, M., Takeoka, Y. 2019. Polyrotaxane actuators. In K. Asaka, H. Okuzaki (eds.) *Soft Actuators*. Singapore: Springer: 81–147.
58. Choi, S., Han, S. I., Jung, D., Hwang, H. J., Lim, C., Bae, S., Park, O. K., Tscabrunn, C. M., Lee, M., Bae, S. Y. et al. 2018. Highly conductive, stretchable and biocompatible Ag–Au core–sheath nanowire composite for wearable and implantable bioelectronics. *Nature Nanotechnology* 13: 1048–1056.

6 Ultra-Thin Graphene Assembly of Liquid Crystal Stretchable Matrix for Thermal and Switchable Sensors

A. P. Meera and P. B. Sreelekshmi

CONTENTS

6.1 INTRODUCTION

Electronic devices such as sensors have already proved their importance in our day-to-day lives. With the advancement of technologies, the need for thermal and electrical appliances is constantly increasing. Among these, stretchable electronics (also known as elastronics) is an emerging and promising area for novel applications such as robotics, stretchable cyber skin, and in vivo implantable sponge-like electronics. The advantage of flexible and stretchable electronics is ease of compression and twisting, which make them good candidates for a variety of applications in health care, energy [1–7], the military, etc.

One of the main challenges is the cost of fabrication, which includes design of materials and novel device configuration. Tireless efforts have been made to realize the stretchability of devices. Another critical challenge is mechanical compliance, in the sense that the device should not undergo any physical damage or change in performance during operation. A number of innovative methods have been developed to overcome these challenges. For making stretchable electronics, elastic materials like rubber could be used, since they can withstand large deformations. But these materials have low electrical conductance. Another alternative is using technology to make non-flexible materials stretchable.

Graphene is an exceptional material used in chemical sensors [7]. Schedin et al. [8] reported that micrometer-size sensors made from graphene are sensitive to adsorption or desorption of gas molecules. To explore the fascinating potential of graphene for a variety of applications, macroscopic ordered structure and effective assembly should be thoroughly investigated.

6.2 LIQUID CRYSTALS

The liquid crystalline state is a thermodynamically stable state of matter with properties in between that of typical solid and liquid phases. True crystalline solids are anisotropic and have an orientational order, while liquids are isotropic and possess fluid properties. Liquid crystals (LCs), often called anisotropic fluids, have structures and properties between anisotropic crystalline solids and isotropic liquids [9–10]. LCs are promising for a variety of applications in display devices, optical devices, and biological sensors [11–14].

Liquid crystalline phases are formed in two ways. One is called lyotropic liquid crystals; water is the solvent, and these are extremely important in biological systems [15, 16]. The second occurs on heating or cooling, and this finds numerous applications in thermal and gas sensors, display devices, etc. [17–19].

Based on the anisotropic structure of mesogens, LC materials can be classified as calamitic (rod like), discotic (disk like) and banana (bent). On the basis of aspect ratio in the anisotropic shape as well as the geometry of the orientation of mesogens, LC phase can be classified into nematic, smectic, and cholesteric [20]. A schematic representation of general LC classification is shown in Figure 6.1.

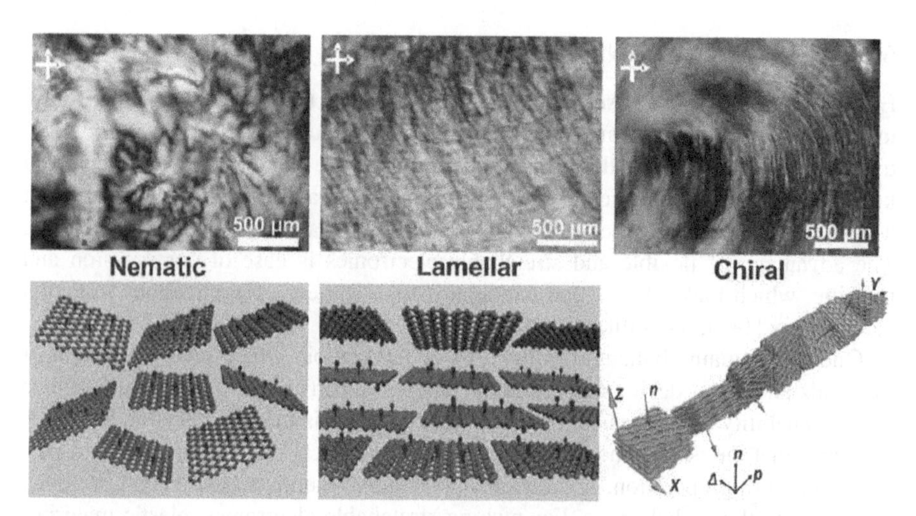

FIGURE 6.1 Typical optical textures of GO liquid crystals (nematic, lamellar, and chiral mesophases) and their corresponding structure models.

Source: Reprinted with permission [24]; copyright 2011 American Chemical Society

In discotic liquid crystals (DLCs), there is a rigid core surrounded by flexible aliphatic chains, and this microphase segregation leads to mesomorphism, as reported by Chandrashakhar et al. [20].

Here, the rigid core behaves like the crystalline phase, and the flexible aliphatic chains behave like the liquid phase. Such discotic behavior has been shown by polycyclic aromatic systems where strong π-π interactions exist. The increase in the number of aromatic systems enhances the π-π overlap, which leads to enhancement of the semiconductor properties of DLCs. Since the structure of graphene is similar to polycyclic aromatic systems, it may be an excellent candidate for electrical and electronic applications as it possesses high charge mobility.

6.2.1 Graphene Oxide Liquid Crystals

Nematic liquid crystal formation for aqueous graphene oxide (GO) dispersion obtained by mild sonication was first reported by Kim at al. [21, 22]. They observed an inhomogeneous, chocolate-like, milky appearance when an aqueous dispersion of low concentration (about 0.05–0.6 wt.%) was kept for a period of three weeks. It was also shown that the upper part is characteristic of the isotropic phase, and the lower dense, chocolate-like part showed optical birefringence.

Conventional methods, such as covalent functionalization, surfactant stabilization, sonication, etc., compromise many desirable properties of graphene and do not control the flake size. The strategy of acid-assisted liquid crystal formation of GO was first developed by Behabtu et al. [23]. They reported the spontaneous exfoliation of graphene and liquid crystalline formation in the presence of chlorosulphonic acid. It was observed that the protonation of graphene by acid could induce repulsion between the layers, and at higher concentrations, the dissolution of graphene requires low levels of protonation.

On increasing the concentration of dispersions, there is a drastic transformation from rigid anisotropic behavior to isotropic liquid crystalline behavior. It was observed that graphene shows this behavior at concentrations up to 20–30 mg/ml. These liquid crystalline phases are promising for functionalization of graphene and nanocomposites. These can also be an alternative for functionalized carbon thin film in various coatings.

In order to obtain liquid crystals of GO, it is important to prepare highly soluble single-layer GO dispersion, which is often challenging. Zhen Xu and Chao Gao reported an isotropic–mematic liquid crystalline phase transition [24]. They prepared a highly soluble single-layer GO dispersion from natural graphite using $KMnO_4$ in concentrated H_2SO_4. There are several advantages for this modified procedure [25], such as strong fluorescence (fourfold), high zeta potential (~ 64 mV), and high transmittance compared to the conventional Hummer's method [26].

The formation of a lyotropic liquid crystal phase was confirmed by polarizing microscopic images (as shown in Figure 6.2), which clearly shows that the isotropic–nematic phase transition starts low at f_m ~2.5×10^{-4}, and on increasing f_m to 5×10^{-3}, the birefringence is shown to spread over the whole dispersion and exhibits a clear schlieren texture, characteristic of nematic phases. It is to be noted that no birefringence is observed for the dispersion with f_m 1×10^{-4} (tube 1). But when f_m is increased to 2.5×10^{-4}, microscopic birefringence develops and a thread-like texture confirms the formation of nematic phase in tube 2, and the textures become more compact from tubes 3 to 7. They also presented a nice schematic model for the isotropic–nematic phase transition. The textures were analyzed by scanning electron microscopy (SEM). From the SEM images, it is evident that the freeze-dried solid characteristic of isotropic dispersion shows disordered clusters (Figure 6.2 (e)) and the one derived

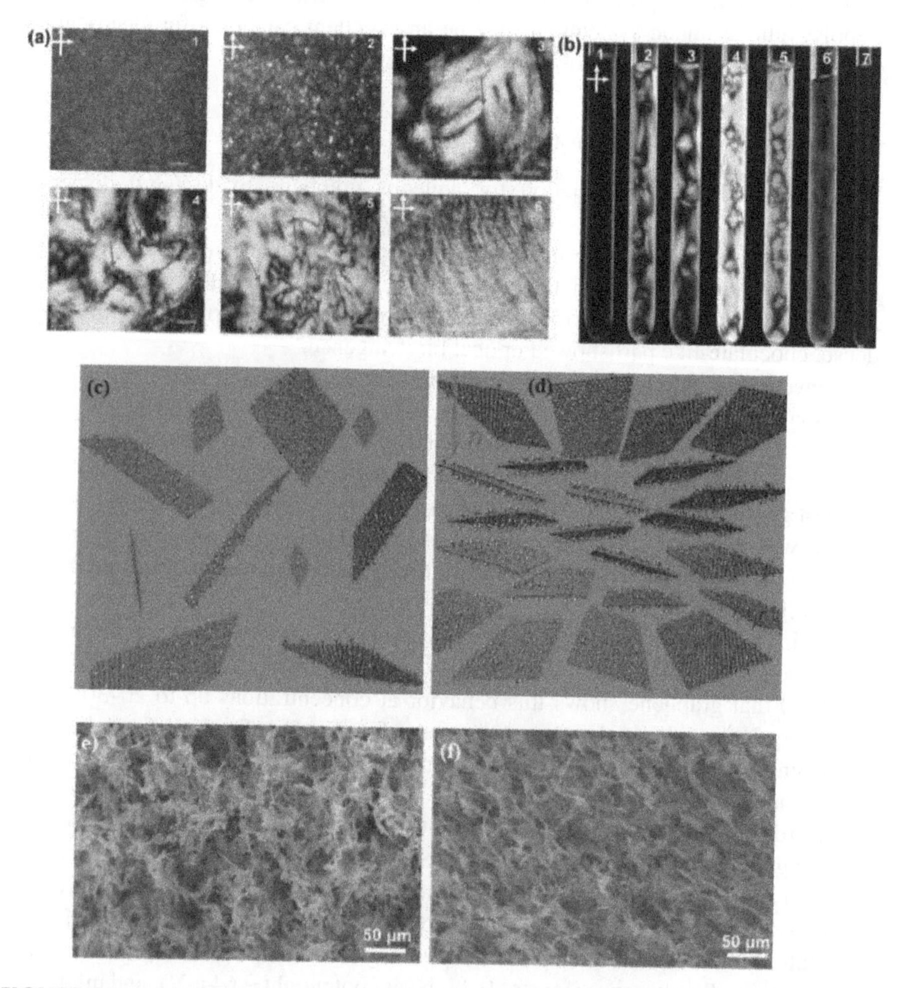

FIGURE 6.2 (a) Polarizing optical microscopic images of GO aqueous dispersions in planar cells between crossed polarizers with f_m's of 5×10^{-4}, 1×10^{-3}, 3×10^{-3}, 5×10^{-3}, 8×10^{-3}, and 1×10^{-2} (from 1 to 6). The arrows indicate the disclinations. (b) GO aqueous dispersions in test tubes with f_m s of 1.0×10^{-4}, 2.5×10^{-4}, 5×10^{-4}, 1.0×10^{-3}, 5×10^{-3}, 1.0×10^{-2} and 2.0×10^{-2} (from 1 to 7). (c) and (d) Schematic models for isotropic and nematic phases, respectively, of GO aqueous dispersion. (e) and (f) SEM images of isotropic dispersion and biphasic system, respectively.

Source: Adapted with permission [24]; copyright 2011, American Chemical Society

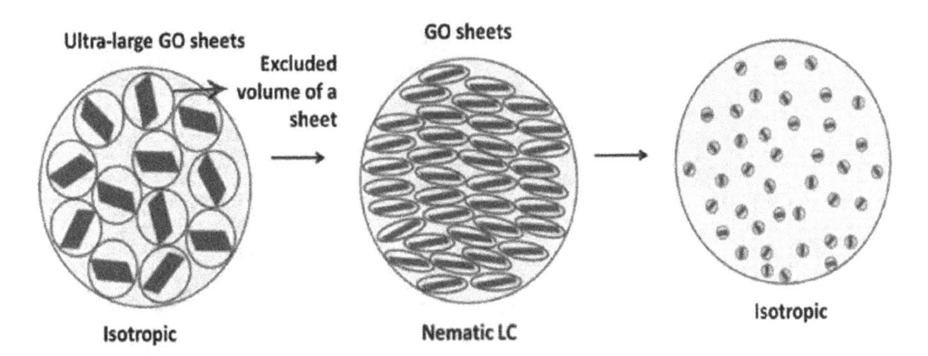

FIGURE 6.3 GO-induced nematic order in GO-isotropic liquid solutions. Left: isotropic phase. Middle: a high enough GO concentration can trigger nematic orientational order. Right: For a low enough aspect ratio characterizing GO, the nematic order cannot form for any concentrations of GO.

Source: Adapted with permission [27]; copyright 2021 Elsevier Ltd

from biphasic phase displays highly oriented domains surrounded by irregular clusters (Figure 6.2 (f)) [24]. As seen from the schematic diagram (Figure 6.3), at low concentrations, the isotropic phase is prominent, and at optimum concentrations, the nematic liquid crystal phase develops [27].

Similar attempts have been made by Guo et al. [28], who tried to prepare single-phase nematic GO by centrifugal vacuum evaporation. They observed a transition from the disordered isotropic liquid phase to an ordered phase with schlieren textures characteristic of nematic liquid crystals at concentrations around 0.5 wt.%. (Figure 6.4). The distribution of GO flake lateral sizes was determined by digital image analysis and found to be in agreement with Onsager model [29] of lyotropic liquid crystals in the disc formulation:

$$\frac{d}{l} \approx 5 \frac{\rho_{\text{GO}}}{\rho_{\text{Suspension}}} C^{-1} \tag{6.1}$$

where d and l are lateral size and thickness of GO sheets, ρ is the density of the materials, and C is the concentration at the isotropic–nematic transition.

The effect of solvents on the liquid crystalline formation was studied by Abedin et al. [30]. As can be seen in Figure 6.5, the lyotropic LC phase formation takes place earlier in water than in other solvents.

6.3 ULTRA-THIN GRAPHENE FILMS

Recently, Li et al. [31] reported a novel, eco-friendly, and cost-effective method of fabricating ultra-thin graphene films (UGFs) that is promising for the manufacture of flexible strain sensors based on the Marangoni effect [32, 33]. The π-π- interactions

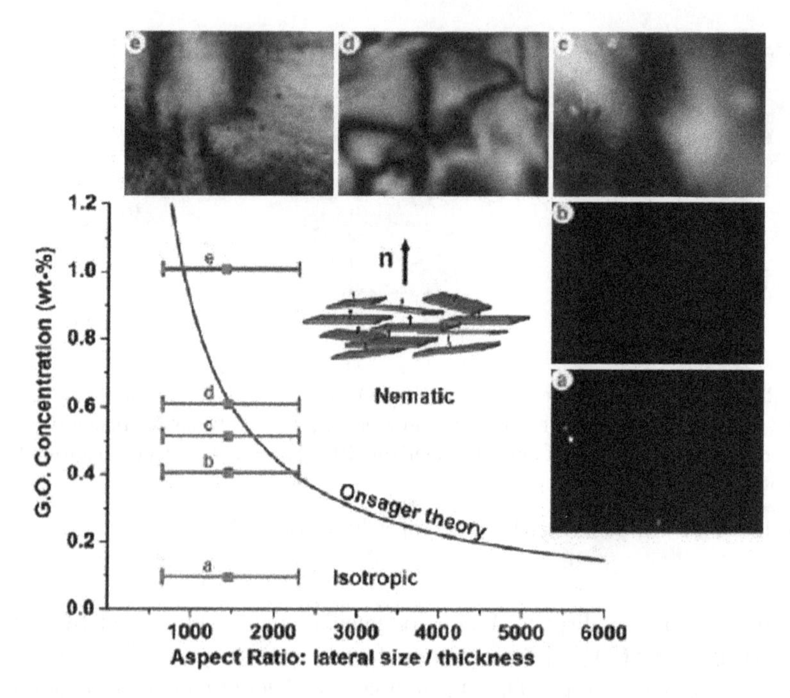

FIGURE 6.4 GO liquid crystal phase diagram. Images are optical micrographs of bulk GO aqueous phases under crossed polarizers. Images show characteristic nematic phase schlieren textures at GO concentrations greater than about 0.4 wt.%. The plot shows the nematic–isotropic transition predicted by the Onsager model.

Source: Adapted with permission [28]; copyright 2011, American Chemical Society

of graphene flakes between two layers form a liquid/air interface. Few-layer graphene flakes were synthesized from natural graphite using the electrochemical exfoliation method. The thickness of graphene flakes was ≈ 2.5 nm, which corresponds to two to three atomic layers. The assembled graphene dispersion possesses high structural regularity compared to other cost-effective solution techniques such as drop-casting and spin-coating [25] (Figure 6.6).

6.4 APPLICATIONS OF GRAPHENE-BASED SENSORS

It was shown that UGFs on the Polydimethylsiloxane substrate could be attached to a human wrist for pulse wave detection and used in monitoring health. Graphene oxide liquid crystals have been used extensively in micromechanical sensors. Due to their high thermal conductivity and light weight, graphene-based structures have been used extensively in thermal management of high-power electronics and portable devices.

Temperature is a physical parameter which plays an important role in monitoring health conditions, artificial electronic skins, and robotics [34–36]. The conventional materials used in temperature sensors have several disadvantages, such as inflexibility, bulkiness, and fragility. Graphene-based materials are promising because of

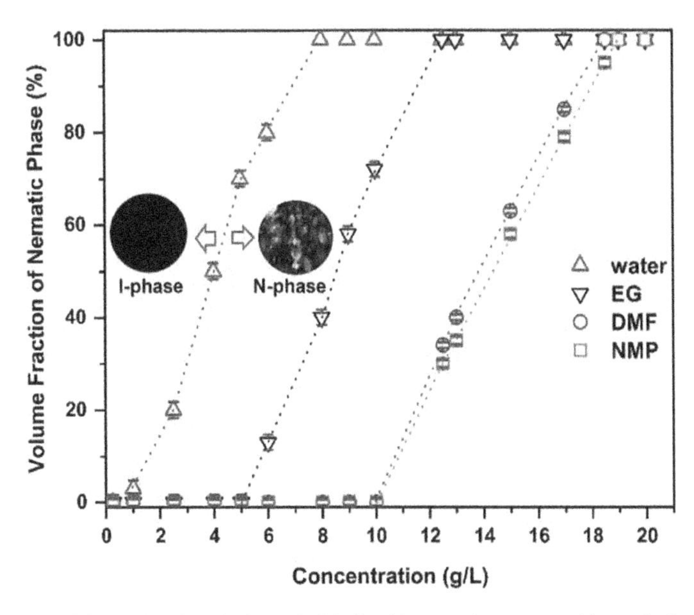

FIGURE 6.5 Orientational ordering of GO liquid crystal. I-N transition of plate-like GO colloids as a function of concentration in different solvents.

Source: Adapted with permission [30]; copyright 2011, American Chemical Society

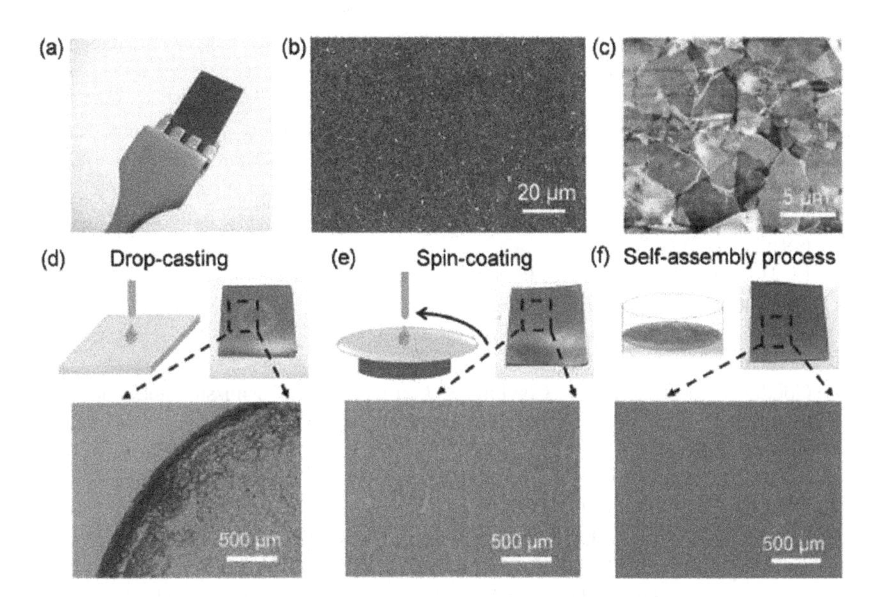

FIGURE 6.6 Characterizations of UGF. (a) Photograph of UGF deposited on an Si substrate. (b) SEM image and (c) atomic force microscopy image of assembled UGF. (d) Drop-casting, (e) spin-coating, and (f) self-assembly process for the production of graphene films.

Source: Adapted with permission [25]

their unique properties. Liu et al. [37] reported a flexible, lightweight, and low-cost temperature sensor based on reduced GO for robot skin.

6.5 CONCLUSIONS

There has been tremendous development in the field of liquid crystals based on graphene in the past few years. Nowadays ultra-thin 2D nanomaterials have emerged as a new class of nanomaterials that researchers can explore to find desirable properties for required applications. For this, well-developed methods may be employed to exfoliate new layered compounds into single- or few-layer nanosheets. The physical, chemical, and electronic properties of ultra-thin 2D nanomaterials are highly dependent on their structural features, which determine their performance. The controlled synthesis of ultra-thin 2D nanomaterials is difficult to materialize. Graphene-based liquid crystals are promising materials for next-generation devices.

REFERENCES

1. Bonaccorso, F., Sun, Z., Hasan, T., Ferrari, A. C. 2010. Graphene photonics and optoelectronics. *Nature Photonics* 4: 611–622.
2. Pang, S., Hernandez, Y., Feng, X., Müllen, K. 2011. Graphene as transparent electrode material for organic electronics. *Advanced Materials* 23(25): 2779–2795.
3. Yong, V., Tour, J. M. 2009. Theoretical efficiency of nanostructured graphene-based photovoltaics. *Small*, 6(2): 313–318.
4. Li, F., Xue, M., Ma, X., Zhang, M., Cao, T. 2011. Facile patterning of reduced graphene oxide film into microelectrode array for highly sensitive sensing. *Analytical Chemistry* 83(16): 6426–6430.
5. Strong, V., Dubin, S., El-Kady, M. F. et al. 2012. Patterning and electronic tuning of laser scribed graphene for flexible all-carbon devices. *ACS Nano* 6(2): 1495–1403.
6. Mukherjee, R., Thomas, A. V., Krishnamurthy, A., Koraktar, N. 2012. Photothermally reduced graphene as high-power anodes for lithium-ion batteries. *ACS Nano* 6(9): 7867–7878.
7. Li, H., Bubeck, C. H. 2013. Photoreduction processes of graphene oxide and related applications. *Macromolecular Research*, 21: 290–297.
8. Schedin, F., Geim, A. K., Morozov, S. V. et al. 2007. Detection of individual gas molecules adsorbed on graphene, *Nature Materials*, 6: 652–655
9. Goodby, J. W., Saez, I. M., Cowling, S. J. et al. 2008. Transmission and amplification of information and properties in nanostructured liquid crystals. *Angewandte Chemie International Edition*, 47(15): 2754–2787.
10. Bisoyi, H. K., Kumar, S. 2011. Carbon-based liquid crystals: Art and science. *Liquid Crystals*, 38(11–12): 1427–1449.
11. Carlton, R. J., Hunter, J. T., Miller, D. S. et al. 2013. Chemical and biological sensing using liquid crystals. *Liquid Crystals Reviews*, 1(1): 1–23.
12. Woltman, S. J., Jay, G. D., Crawford, G. P. 2007. Liquid-crystal materials find a new order in biomedical applications. *Nature Materials* 6: 929–938.
13. Jalili, R., Aboutalebi, S. H., Esrafilzadeh, D. et al. 2013. Scalable one-step wet-spinning of graphene fibers and yarns from liquid crystalline dispersions of graphene oxide: Towards multifunctional textiles. *Advanced Functional Materials*, 23(43): 5345–5354.

14. Xiang, C., Young, C. C., Wang, X. et al. 2013. Large flake graphene oxide fibers with unconventional 100% knot efficiency and highly aligned small flake graphene oxide fibers. *Advanced Materials*, 25(33): 4592–4597.
15. Hamley, I. W. 2010. Liquid crystal phase formation by biopolymers. *Soft Matter*, 9: 1863–1871.
16. Nakata, M., Zanchetta, G., Chapman, B. D. et al. 2007. End-to-end stacking and liquid crystal condensation of 6 to 20 base pair DNA duplexes. *Science*, 318(5854): 1276–1279.
17. Geelhaar, T., Griesar, K., Reckmann, B. 2013. 125 years of liquid crystals–a scientific revolution in the home. *Angewandte Chemie International Ed*ition, 52(34): 8798–8809.
18. Bremer, M., Kirsch, P., Memmer, M. K., Tarumi, K. 2013. The TV in your pocket: Development of liquid-crystal materials for the new millennium. *Angewandte Chemie International Ed*ition, 52(34): 8880–8896.
19. Boden, N., Clements, J., Movaghar, B. 2002. Fluid sensing device using discotic liquid crystals. *US006423272B1*.
20. Chandrasekhar, S., Sadashiva, B. K., Suresh, K. A. 1977. Liquid crystals of disc-like molecules. *Pramana*, 9: 471–480.
21. Kim, S. O., Kim, J. E., Han, T. H., Lee, S. H., Kim, J. Y. 2013. Korea Advanced Institute of Science and Technology, *US8449791 B2*.
22. Kim, S. O., Kim, J. E., Han, T. H., Lee, S. H., Kim, J. Y. 2012. Korea Advanced Institute of Science and Technology, *KR 10–1210513*.
23. Behabtu, N., Lomeda, J. R., Green M. J. et al. 2010. Spontaneous high-concentration dispersions and liquid crystals of graphene. *Nature Nanotechnology*, 5: 406–411.
24. Xu, Z., Gao, C. 2011. Aqueous liquid crystals of graphene oxide. *ACS Nano*, 5(4): 2908–2915.
25. Marcano, D. C., Kosynkin, D. V., Berlin, J. M. et al. 2010. Improved synthesis of graphene oxide. *ACS Nano*, 4(8): 4806–4814.
26. Hummers, W. S., Offeman R. E. Preparation of graphite oxide. *Journal of the American Chemical Society*, 1958, 80(6): 1339.
27. Pal, K., Aljabali, A. A., Kralj, S., Thomas, S., Souza, F. G. 2021. Graphene-assembly liquid crystalline and nanopolymer hybridization: A review on switchable device implementations. *Chemosphere* 263: 128104.
28. Guo, F., Kim, F., Han, T. H., Shenoy, V. B., Huang, J., Hurt, R. H. 2011. Hydration-responsive folding and unfolding in graphene oxide liquid crystal phases. *ACS Nano*, 5(10): 8019–8025.
29. Forsyth, P. A., Marcelja, S., Mitchell, D. J., Ninham, B. W. 1977. Onsager transition in hard plate fluid. *Journal of the Chemical Society, Faraday Transactions 2: Molecular and Chemical Physics* 1: 84–88.
30. Abedin, M. J., Gamot, T.D., Martin, S. T. et al. 2019. Graphene oxide liquid crystal domains: Quantification and role in tailoring viscoelastic behavior. *ACS Nano*, 13(8): 8957–8969.
31. Li, X., Yang, T., Yang, Y. et al. 2016. Large-area ultrathin graphene films by single-step Marangoni self-assembly for highly sensitive strain sensing application. *Advanced Functional Materials*, 26(9):1322–1329.
32. Huang, J., Kim, F., Tao, A. R., Connor, S., Yang, P. 2005. Spontaneous formation of nanoparticle stripe patterns through dewetting. *Nature Materials*, 4, 896–900.
33. Shim, J., Yun, J. M., Yun, T. 2014. Two-minute assembly of pristine large-area graphene based films. *Nano Letters*, 14(3): 1388–1393.
34. Gao, L., Zhang, Y., Malyarchuk, V. et al. 2014. Epidermal photonic devices for quantitative imaging of temperature and thermal transport characteristics of the skin. *Nature Communications*, 5: 4938.

35. Kanao, K., Harada, S., Yamamoto, Y. et al. 2015. Highly selective flexible tactile strain and temperature sensors against substrate bending for an artificial skin. *RSC Advances*, 38: 30170–30174.
36. Choong, C. L., Shim, M. B., Lee, B. S. et al. 2014. Highly stretchable resistive pressure sensors using a conductive elastomeric composite on a micropyramid array. *Advanced Materials*, 26(21): 3451–3458.
37. Guanyu, L., Qiulin, T., Hairong, K. et al. 2018. A flexible temperature sensor based on reduced graphene oxide for robot skin used in Internet of Things. *Sensors*, 18(5): 1400.

7 CNT/Graphene-Assisted Flexible Thin-Film Preparation for Stretchable Electronics and Superconductors

Vinayak Adimule, Santosh S. Nandi, B. C. Yallur, and Nilophar Shaikh

CONTENTS

7.1 INTRODUCTION

The widespread applications of stretchable electronic devices due to easier processibility, low-cost production, and reliability have attracted much interest in the fields of robotics, flexible devices, superconductors, and conformable materials [1–4]. Wearable and implantable soft electronic devices find most applications because of their high conductivity and ability to be mechanically deformable [5, 6]. However the brittle and rigid form of the bulk material used in conventional electronic devices is being replaced by soft electronic materials, and many such materials have been used, including inorganic nanostructured materials [7], organic and inorganic hybrid materials [8], bimetallic oxides [9], and 2D nanomaterials [10]. Among many such materials, CNTs have attracted much attention due to good processibility, high conductivity, and high intrinsic carrier mobility [11, 12]. The high quality of single-walled carbon nanotubes (SWCNTs) can be synthesized in a variety of ways, including arch discharge [13], laser ablation [14], and chemical vapor deposition [15]. The covalent sp^2 hybridization that exists in CNTs means they possess high stiffness and flexibility [16]. Young's modulus of CNTs, investigated by atomic force microscopy and transmission electron microscopy, lies between 270 GPa and 970 GPa [17, 18]. Graphene and CNTs can be transformed with the application of pressure. Commonly in CNT networks, conductivity is due to the conductive channels formed by CNTs, tunnelling between neighboring CNTs, and resistance offered by the CNTs plays vital role in the piezoelectrical properties and conductivity of the CNTs [19]. An experiment carried out by Hu et al. found a threshold limit of current conductivity of 0.1 wt.% polymer/CNT composite, and above 2 wt.%, electrical conductivity reaches saturation level due to percolation [20]. Fabrication of CNTs depends largely on the viscosity gradient and percolation. The relationship between two has been studied by Bauhofer and Kovacs [21]. Particular alignment of CNTs and CNT composite rendered conductivity in the order of 1,000 S/m for 2 wt.% CNT in polystyrene. Avilés et al. designed multi-walled carbon nanotubes (MWCNTs) in an alternating electric field; application of 6 kV/m and 7.3 kV/m showed linear conductivity over the whole range of the experimental determination [22]. Low CNT concentration (0.1 wt.% to 0.5 wt.%) of the composite is five orders of magnitude greater than the randomly distributed CNT. In general, CNTs have mixture of chirality, diameter, and defects with high surface area, and they form bundles and aggregate themselves with the help of van der Waals forces between them [23]. However, intensive mechanical processing changes structure the CNTs and introduce more defects.

In the present investigation of the superconductivity of CNTs, we discuss in detail the mechanism of transformation from graphene, the electrical conductivity of the various CNTs and composites, their flexibility and their application in soft electronic materials.

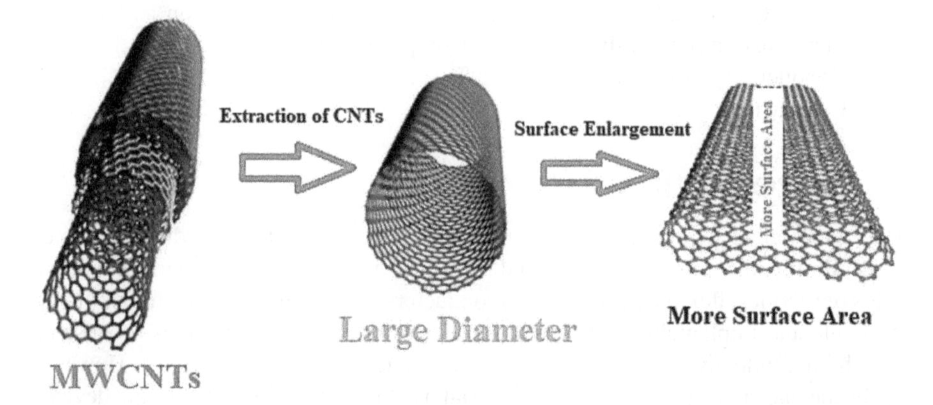

FIGURE 7.1 Schematic illustration of MWCNTs and formation of large area of CNT network.

7.2 SYNTHESIS OF CNTS (SOLUTION-BASED METHOD) AND FABRICATION OF THIN FILMS

Lorem CNTs can be synthesized and fabricated by a variety of methods, such as drop coating [24], printing [25, 26], spin coating [27], and solution evaporation [28]. On dropping CNT solution on a spinning substrate [29], the deposited CNTs align in different directions, locating on the substrate with wafer-scale uniformity. Drop coating and printing fabrication methods are more promising for scalability of CNT networks during the thin film preparation of the transistors. In order to achieve high uniformity of the CNT network in the wafer, the substrate is immersed in the amine-containing molecule. This enables SMCNTs to form thin film transistors for incorporation in the fabrication of the device [30, 31]. The printing techniques are a low-cost method for large-scale fabrication of SWCNT thin film transistors and circuits or gravure printing [32]. Resolutions obtained from the printing process are lower than those from conventional printing and provide large-surface-area, cost-effective CNT circuits with moderate performance in devices. During this process, high-performance soft electronic circuits made of the thin films of CNT networks are obtained.

7.3 SCANNING ELECTRON MICROSCOPY CHARACTERIZATION

Scanning electron microscopy (SEM) characterization carried out in order to understand the morphologies of CNT networks doped with 0.5 wt.% to 2 wt.% are shown in Figure 7.2. Figure 7.2 (a) represents the CNT-fabricated thin film transistor with CNT nanoparticles (NPs) that have an average size of ~ 85 nm, interpore diameter of ~ 40 nm, and CNT length of ~ 50 μm. Figure 7.2 (b) shows how the NPs developed as a result of the solution evaporation method, with average NP size of ~ 110 nm, interpore diameter of ~ 38 nm, and CNT length of ~ 38 nm. Figure 7.2 (c) and (d) reveal dense agglomeration of the CNT NPs dispersed over the substrate underlying thc area with average size of NPs varying between ~ 78 nm and ~ 115 nm, with a length of ~ 55 μm.

7.4 FIELD EFFECT TRANSISTOR FABRICATION METHODS

Fabrication of CNT FETs depends on alignment of the CNTs, and extensive research has been done to overcome limitations such as inconsistency in device performance, poor logic outputs, and increased power consumption [33]. In order to minimize inconsistency and to maximize performance, CNT networks are deposited over a large area of the substrate surface with high uniformity so as to achieve large current density when low voltage is applied. Channel dimension is essential to achieving optimal current density; to achieve low power density on a flexible substrate, a scalable approach was developed by Xiang et al. [34]. There are diverse applications for CNTs deposited uniformly over flexible substrate, especially in the field of NOT-AND/NOT-OR logic gates, inverters, and integrated circuits, among others. In order to achieve visual pressure information in real time, Wang et al. [35] developed CNT field effect transistor (FET) device circuitry with monolithically integrated pressure

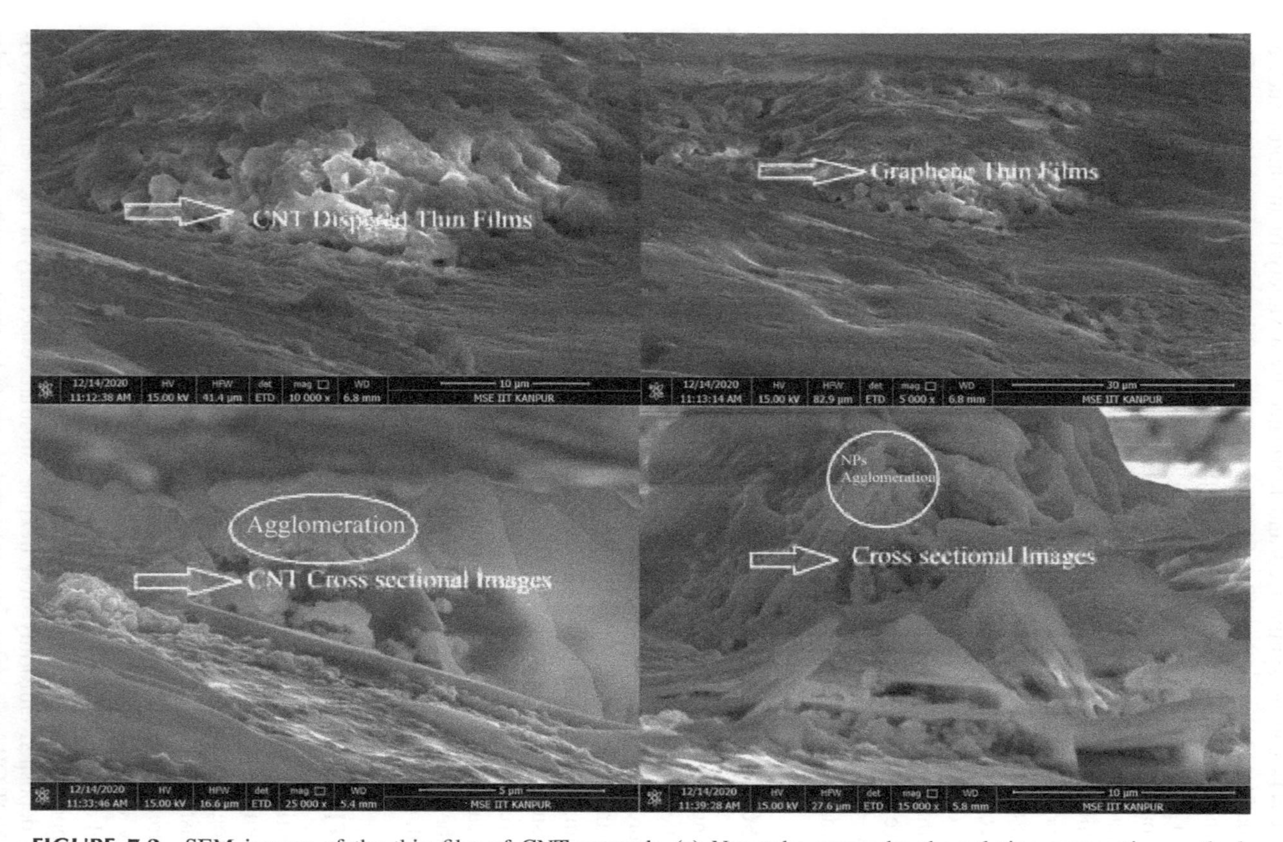

FIGURE 7.2 SEM images of the thin film of CNT network. (a) Nanotube grown by the solution evaporation method. (b) Fabricated SEM micrographic image. (c) Cross-sectional CNT SEM image of the thin film. (d) CNT NPs agglomerated thin film transistor.

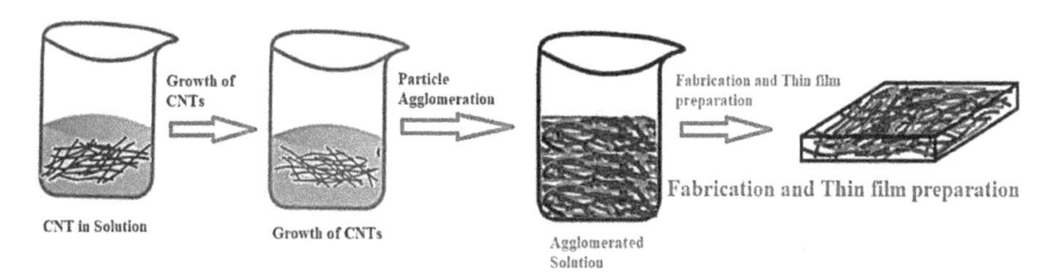

FIGURE 7.3 Schematic representation of CNT solution processing for thin film preparation.

sensors and organic light-emitting diodes. To maximize the mechanical stability of flexible devices, Sekiguchi et al. [36] demonstrated wafer-scale fabrication of CNT FET devices. Such devices consisted of elastomeric electrodes with metallic electrodes and ion-gel-based gate dielectrics that exhibited exceptional mechanical properties. This work demonstrates the feasibility of introducing CNT FETs in the making of flexible electronic devices.

7.5 SOFT ELECTRONICS (CNTS) AND INTRINSICALLY STRETCHABLE ELECTRONICS

The mechanical elasticity of CNT nanotubes means they are easily used as intrinsically stretchable electronic components, such as gate dielectrics; active semiconducting channels and intrinsic diodes maintain high-quality interfaces between the electrical components attached to the substrate surface. Such devices find applications as implantable or wearable biosensors; for example, confocal contacts [37, 38]. CNT FETs were deposited on a glass substrate to form a network of thin films and transferred to a 50% PDMS surface creating buckled CNT FET networks. No cracks were developed on a CNT surface even when the PDMS substrate was re-stretched to 65–70%. Releasing strain makes the electrodes stretchable. Chortas et al. experimentally determined CNT FET that was stretchable even at higher degrees of strain (up to 100%) [39]. Metallic CNTs/SWCNTs were spray coated on a Si wafer substrate and subjected to photolithography to form gate, source, and drain terminals.

7.6 CNT THIN FILM AS STRAIN SENSORS

In CNTs, strain sensitivity has a threshold (high strain sensitivity) due to percolation [40]. The percolation threshold depends on factors like type of CNT, their aspect ratio, dispersion quality, and functionalization of the CNT network. Usually, functionalized CNT networks have a higher threshold percolation than non-functionalized CNT networks. Percolation threshold and conductivity depend on many factors, like aspect ratio, disentanglement of CNT agglomerates, degree of spatial orientation of the CNT network, etc. [41]. Addition of CNT to the polymer and composite mixture significantly improves the conductivity of the polymer. Critical exponent (t) depends on the percolation threshold (θc) and weight fraction of the conducting CNT surface. According to the theory, $t = 1.6$ for a 2D CNT network and $t = 2$ for a 3D CNT network.

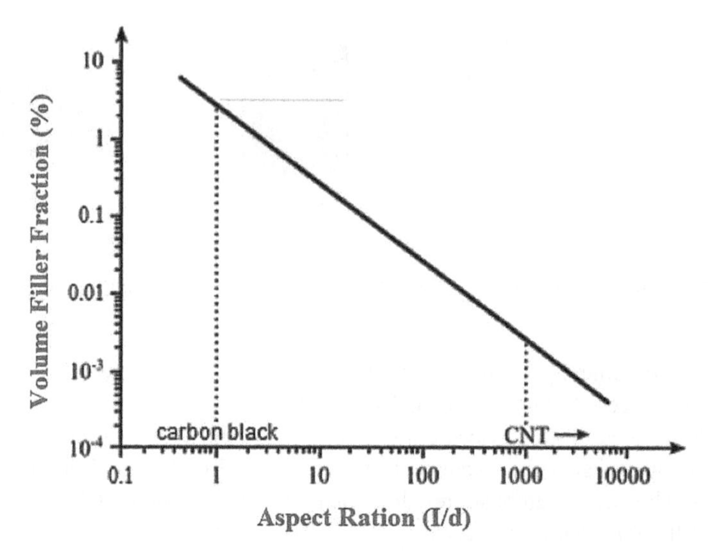

FIGURE 7.4 CNT network graph showing volume filler loaded fractions vs aspect ratio.

7.7 CNT THIN FILM AS SUPERCONDUCTORS

7.7.1 Electrical Conductivity and Piezoelectric Characterization

Conductivity of a 2 wt.% CNT network reaches saturation level, as shown in Figure 7.5. Viscosity of the CNT material plays a very important role in the fabrication process. The relational ship between conductivity and viscosity of the CNT was reported by Bauhofer et al. [42]. Conductivity in the order of 1,000 S/m can be obtained for a CNT network dispersed in polystyrene molecule. For SWCNTs/MWCNTs which are randomly distributed, conductivity was lower than when they are dispersed regularly. The relationship between conductivity and viscosity of the MWCNTs over a whole range is illustrated in Figure 7.6. Typical piezoelectrical conductivity of the aligned CNT network and variation in the resistance of the material is shown in Figure 7.7. For 0.5 wt. % and 2 wt. %, variation in the total resistance was linear. Two types of resistance values are determined for the aligned bundles of CNT network.

Intrinsic resistance is denoted by R_{tube}, and intertube resistance is denoted by $R_{junction}$, which varies from 0.2 KΩ s/μm to 0.4 KΩ s/μm. The total resistance is given by the following equation:

$$R = R_{tube} + R_{junction} \quad 7.2$$

Piezoresistivity/electrical conductivity is considerably affected by variation in conductivity of the CNT network, deformation in the CNT, and distance between neighboring CNT networks. Hu et al. [43] effectively explained the piezoresistivity of the CNT using the 3D fiber reorientation model. They found that resistance of CNT strains (having a concentration less than 1 wt.%) which are close to threshold

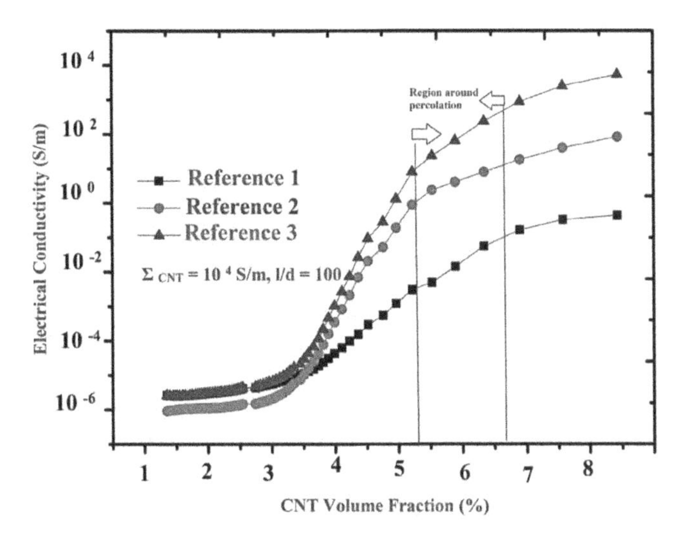

FIGURE 7.5 Relationship between electrical conductivity and CNT volume fractions.

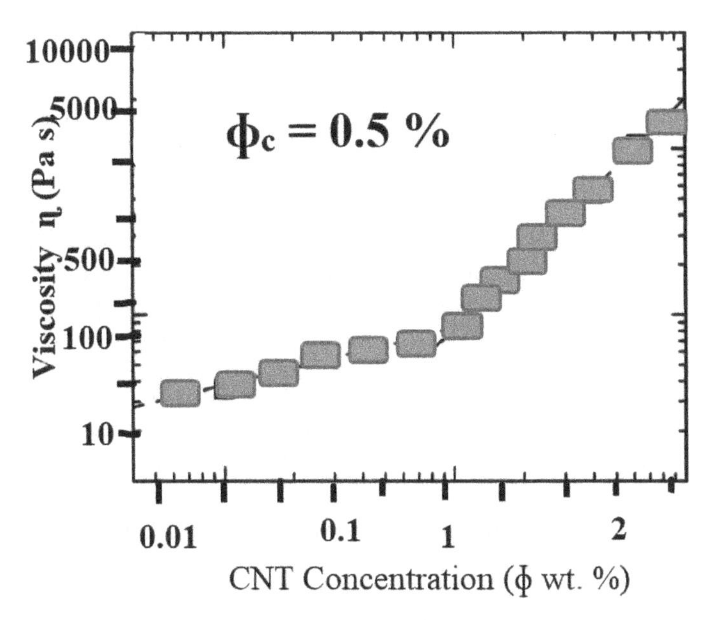

FIGURE 7.6 CNT network with the relationship between viscosity of the material and CNT concentration.

percolation causes a tunneling effect. The piezoresistivity of the material can be calculated by using the gauge factor (K), given by the following equation:

$$K = 1/dl \ dR/R \tag{7.2}$$

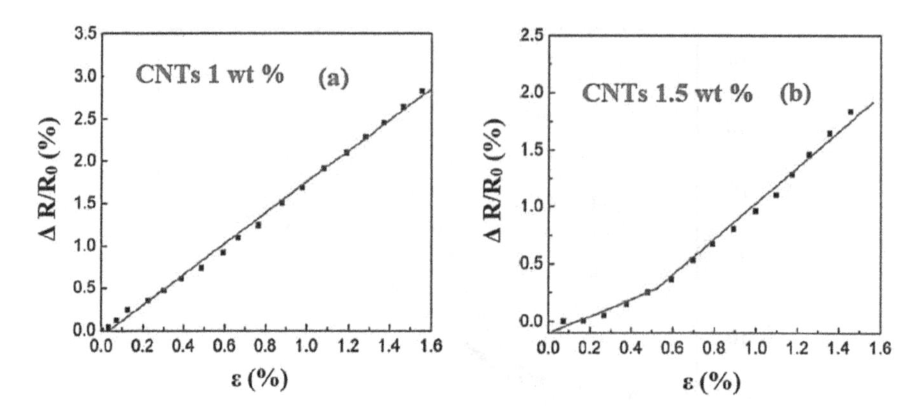

FIGURE 7.7 Variation of piezoresistivity with respect to conductivity of the CNT network.

where K is gauge factor, dR/R is relative change in the resistance of the CNT network, and l/dl = ε – strain factor.

7.8 SUPERCONDUCTIVITY OF THE CNT/CNT COMPOSITE NETWORK

Commonly, a polymer-doped CNT nanocomposite exhibits a lower conductivity (k) value than a polymer composed of an MWCNT network. The A polymer composed of an MWCNT network synthesized by the chemical vapor deposition method shows a minimum k value of 0.034 W/MK compared to a pure CNT network [44]. Thus, in a polymer composed of CNT, graphene can produce competent, low-cost, flexible conducting composites for next-generation devices and applications. Moreover, the influence of stabilizers in SWCNTs or MWCNTs on electron transport properties of the polymer-doped composite has been shown by Kim et al [45]. in such polymer composites, electron transport will be greatly influenced by modification of the CNT tube junction in the entire network. Conductivity and resistivity values are unaltered for 40 wt.% of SWCNT observed with 900 S/m.

Generally, Au particles are introduced into the CNT/SWCNT/MWCNT nanomatrix in order to increase the conductivity of the composites [46] the carrier concentration increases upon doping. The effect of Au concentration on the CNT matrix and its conductivity parameters have also been studied. Choi et al. has studied σ of the composite and conductivity (S) of the composite; generally, these are enhanced by 110–115 S/m, and S of the composites became slightly insensitive to the Au dopant concentration.

In a recent investigation, several metal NPs were also shown to enhance S rather than σ. The commonly used PEDOT:PSS enhances the S value and possesses several other features like high conductivity, optical transference, high flexibility, thermal conductivity, and ease of use. A PEDOT:PSS mixture may be a promising material for thermoelectric material applications. A limitation of using PEDOT:PSS was its solubility in aqueous media. Properties of the SWCNT were enhanced by the filler loadings, and it was also found that the σ value obtained from the experiment was much higher than the calculated σ value. A novel chemical synthetic method was used by Kim et al. [47] in experimental work on a Polyaniline/CNT/nanohybrid material; this was activated by the addition of ammonium peroxydisulphate into the CNT matrix. They also reported the various superconductivity properties of the polymer hybrid material with respect to CNT concentration. In addition, they discusses single-walled CNTs, various oxidized and unoxidized CNTs, and alternative hybrid materials which show superior properties compared to PANI/CNT nanomaterials.

7.9 CONDUCTIVITY OF POLYMER-DOPED CNT

Large molecular weights of polyethylene/CNT (MWCNT), epoxy/CNT (MWCNT), poly(ethylene terephthalate) (PET)/CNT (MWCNT), epoxy/CNT (MWCNT) [48–50] were chosen for experimental investigation of their relative superconductivity. A percolation threshold was applied for all the polymer-dispersed CNT material and vales of *S* obtained were 950, 800, 850, and 657 S/m, respectively. The superconductivity values of the polymer-dispersed MWCNT material is presented in Table 7.1.

The output values demonstrate the behavior of the MWCNT material and the interfacial conductivity in samples reported in the literature. The highest value of interfacial conductivity was seen in case of a polyethylene-dispersed MWCNT network, and poly(ethylene terephthalate) (PET)/CNT (MWCNT) also showed good interfacial conductivity compared to a polyethylene molecule. The conductivity calculations revealed the formation of different interphase regions of the polymer-doped CNT framework, and the thickest interface was observed for the polyethylene-dispersed MWCNT samples. Weakest conductivity was reported for the epoxy-dispersed CNT network; this is due to transference from the CNT NPs to the epoxy polymeric material. Figure 7.8 shows the conductivity measurement curves for the samples under investigation.

The interfacial conductivity and interfacial regions affect the conductivity of the doped and undoped polymeric CNT materials under investigation. It is clear that the predictions from the experimental values are closer to the ones reported in the literature. The model that was developed can be used to estimate the applied conductivity of the doped and undoped CNT networks. The interfacial conductivity, interfacial

TABLE 7.1
Superconductivity measurements obtained from resistivity vs different weight of CNT and polymer-doped CNT materials

Sample	Tc ^{off} (K)	Tc 1 ^{on} (K)	Tc 2 ^{on} (K)	ΔTc 1(K)	σ^0(μΩ-cm)	T*K
Pure CNT	88.95	84.56	-	4.30	3,052.8	74.12
CNT/polyethylene	92.14	84.14	-	8.00	3,214.8	75.47
CNT/epoxy	93.45	82.47	-	10.98	3,147.8	71.24
CNT/PET	97.18	82.33	-	14.85	3,112.8	72.45
MWCNT (0.5 at wt.%)	98.12	81.97	-	16.15	3,478.9	70.14
MWCNT (0.75 wt.%)	95.64	80.74	-	14.90	3,458.9	72.94
MWCNT (1 wt.%)	93.24	80.54	-	12.70	3,012.4	76.94

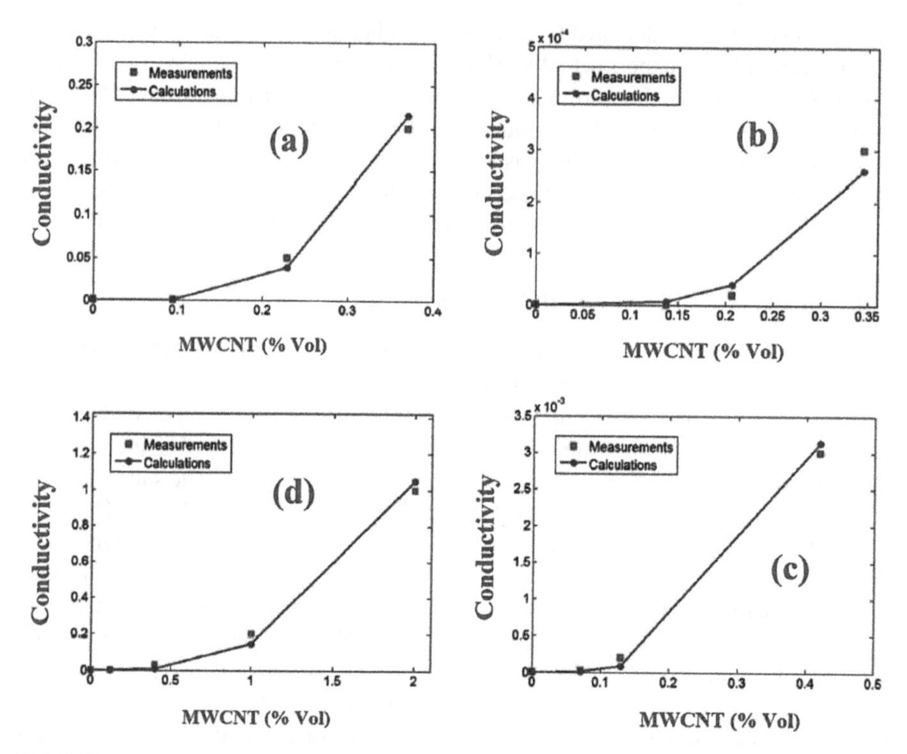

FIGURE 7.8 Conductivity measurements curves for different wt.% of MWCNT at various concentrations: (a) 0.25 wt.%, (b) 0.75 wt.%, (c) 1.15 wt.%, and (d) 1.65 wt.% of polyethylene molecule.

regions and tunneling effect can replace conventional models. The dissimilar tunneling distances induce different resistivity and conductivity and enables the charge transferring process to occur, which in turn affects the nanocomposite materials and their performance as a superconducting material [51, 52].

The impacts of parameters on the conductivity of polymeric composite material was also investigated for the various doped materials and their conductivity of the nanocomposite material. Figure 7.9 displays the values for resistivity (*R*) and conductivity, and the average values of the other parameters are shown in Table 7.2. The conductivity maximum decreases as the R value decreases. Accordingly, the thin film of the membrane and conducting CNT framework improve the conductivity of the device. The thin film of the CNT also enhances the size and density of the polymer-doped nanonetwork and the MWCNT, and with increase in the length and concentration of CNT in polymeric material, conductivity also increases. Table 7.2 gives the crossover potential, superconductivity coherence, and Josephson coupling constant. However, conductivity of the MWCNT depends directly on the conductivity of the CNT filler. A comparison of various properties of graphene molecules and CNT is provided in Table 7.3.

TABLE 7.2
Values depicting excess superconductivity measurements for CNT/graphene networks

Sample	Crossover temperature T_{LD} (K)	Superconductivity coherence length (ϕ)	Johnson coupling constant (E_j)
Pure CNT	94.11	1.012	0.24
CNT/polyethylene	97.84	1.845	0.98
CNT/epoxy	95.14	1.652	0.45
CNT/PET	99.54	1.847	0.54
MWCNT (0.5 wt.%)	96.47	1.257	0.64
MWCNT (0.75 wt.%)	97.85	1.354	0.78
MWCNT (1 wt.%)	98.99	1.247	0.47

TABLE 7.3
Comparison of properties of graphene and CNT network

Property	Unit	Graphene	CNT network
Fracture strength	GPa	124	45
Density	g/cm^3	>1	1.33
Thermal conductivity	W/m K	5000	3000
Electrical conductivity	S/cm	10^6	5000
Charge mobility	$cm^2/V/s$	2000000	100000

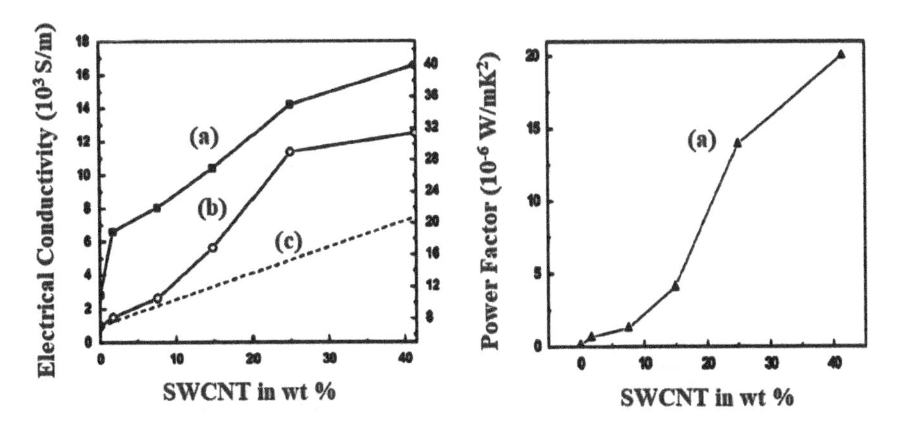

FIGURE 7.9 Electrical conductivity measurements for SWCNT and overall power factor for different wt.% of SWCNT material at room temperature.

7.10 CONCLUSION

In summary, in the present investigation, superconductivity of the CNT/graphene network system was discussed along with their polymer-doped synthesis, properties, and electrical conductivity parameters. Interfacial conductivity and the CNT network were characterized by percolation, tunneling effect, threshold percolation, etc. Thick interference in the conductivity was observed for the thin film of a polymer-doped CNT network. Moreover, in a long CNT polymer-doped network, a thick interface in the MWCNT decreased the percolation threshold of the polymer network and increased the conductivity. Additionally the experimental procedure that was developed shows good agreement with the results obtained from experiments. So the developed model can be a better harvester for tunneling conductivity, interfacial region, and interfacial conductivity. The experiment measured superconductivity of the different wt.% of polymer-doped CNT networks and the different wt.% of MWCNTs/SWCNTs. The MWCNT network at 0.5 wt.% showed superconductivity of 3478.9 μΩ-cm, and this decreased as the concentration of MWCNT increased. For the polymer-doped CNT network, a polyethylene molecule showed highest conductivity: 3214.8 μΩ-cm at 298 K. The authors also made various calculations based on an experiment of superconductivity; these were excess coherence, Josephson coefficients, and crossover potentials for various CNT-doped polymers as well as MWCNT/CNT/graphene network systems.

The addition of CNT/graphene systems to various polymers developed pinning centers inside the polymeric matrices. For 1 wt.% to 5 wt. %, superconductivity decreases as a result of excess coherence and crossover potentials. Appropriate methodologies and techniques are still to be developed to improve the percolation threshold, tunneling conductivity, interfacial region, and interfacial conductivity of CNT/graphene networks.

ACKNOWLEDGMENTS

All authors are thankful to MSRIT for SEM, fabrication, and superconductivity measurements. They are thankful also to CeNSE, IISc, Bangalore, for electrical conductivity, I-V, C-V measurements, and investigation.

REFERENCES

1. Rogers, J.A., Someya, T., Huang, Y. 2010. Materials and mechanics for stretchable electronics. *Science* 327:1603–1607.
2. Someya, T., Kato, Y., Sekitani, T., Iba, S., Noguchi, Y., Murase, Y., Kawaguchi, H., Sakurai, T. 2005. Conformable, flexible, large-area networks of pressure and thermal sensors with organic transistor active matrixes. *Proceedings of the National Academy of Science* 102(35):12321–12325.
3. Ahn, J.H., Je, J.H. 2012. Stretchable electronics: Materials, architectures and integrations. *Journal of Physics D: Applied Physics* 45(10):103001.
4. Kim, D.H., Rogers, J.H. 2008. Stretchable electronics: Materials strategies and devices. *Advanced Materials* 20(24):4887–4892.
5. Jaemin, K., Ghaffari, R., Kim, D.H. 2017. The quest for miniaturized soft bioelectronic devices. *Nature Biomedical Engineering* 1:0049.

6. Sungmook, J., Jongsu, L., Taeghwan, H., Minbaek, L., Dae-Hyeong, K. 2014. Fabric-based integrated energy devices for wearable activity monitors. *Advanced Materials* 26(36):6329–6334.
7. Viventi, J., Kim, D.H., Moss, J.D., Kim, Y.S., Blanco, J.A., Annetta, N., Hicks, A., Xiao, J., Huang, Y., Callans, D.J., Rogers, J.A., Litt, B. 2010. A conformal, bio-interfaced class of silicon electronics for mapping cardiac electrophysiology. *Science Translational Medicine* 2(24):24–22.
8. Banger, K.K., Yamashita, Y., Mori, K., Peterson, R.L., Leedham, T., Rickard, J., Sirringhaus, H. 2011. Low-temperature, high-performance solution-processed metal oxide thin-film transistors formed by a sol–gel on chip process. *Nature Materials* 10:45–50.
9. Chang, H.Y., Yang, S., Lee, J., Tao, L., Hwang, W.S., Jena, D., Lu, N., Akinwande, D. 2013 High-performance, highly bendable MoS_2 transistors with high-K dielectrics for flexible low-power systems. *ACS Nano* 7(6):5446–5452.
10. Son, D., Chae, S.I., Kim, M., Choi, M.K., Yang, J., Park, K., Kale, V.S., Koo, J.H., Choi, C., Lee, M., Kim, J.H., Hyeon, T., Kim, D.H. 2016. Colloidal synthesis of uniform-sized molybdenum disulfide nanosheets for wafer-scale flexible nonvolatile memory. *Advanced Materials* 28(42):9326–9332.
11. Jang, S., Jang, H., Lee, Y., Suh, D., Baik, S., Hong, B.H., Ahn, J.H. 2010. Flexible, transparent single-walled carbon nanotube transistors with graphene electrodes. *Nanotechnology*, 21(42):425201.
12. Gamaly, E.G., Ebbesen, T.W. 1995. Mechanism of carbon nanotube formation in the arc discharge. *Physical Review B* 52:2083.
13. Scott, C.D., Arepalli, S., Nikolaev, P., Smalley, R.E. 2001. Growth mechanisms for single-wall carbon nanotubes in a laser-ablation process. *Applied Physics A* 72:573–580.
14. Che, G., Lakshmi, B.B., Martin, C.R., Fisher, E.R. 1998. Chemical vapor deposition-based synthesis of carbon nanotubes and nanofibers using a template method. *Chemistry of Materials* 10(1):260–267.
15. Nguyen, V.C., Cao, T.T., Nguyen, V.T., Vuong, T.Q., Vuong, T.Q.P., Pham, V.T., Ngo, T.T.T. 2015. A simple approach to the fabrication of graphene-carbon nanotube hybrid films on copper substrate by chemical vapor deposition. *Journal of Materials Science & Technology* 31(5):479–483.
16. Dresselhaus, M.S., Dresselhaus, G., Eklund, P.C. 1996. *Science of Fullerenes and Carbon Nanotubes: Their Properties and Applications*. New York: Academic Press.
17. Yu, M.F., Lourie, O., Dyer, M.J., Moloni, K., Kelly, T.F., Ruoff, R.S. 2000. Strength and breaking mechanism of multiwalled carbon nanotubes under tensile load. *Science* 287(5434):637–640.
18. Ruoff, R.S., Tersoff, J., Lorents, D.C., Subramoney, S., Chan, B. 1993. Radial deformation of carbon nanotubes by van der Waals forces. *Nature* 364:514–516.
19. Hu, N., Karube, Y., Arai, M., Watanabe, T., Yan, C., Li, Y., Liu, Y.L., Fukunaga, H. 2010. Investigation on sensitivity of a polymer/carbon nanotube composite strain sensor. *Carbon* 48(3):680–687.
20. Hu, N., Masuda, Z., Yan, C., Yamamoto, G., Fukunaga, H., Hashida, T. 2008. Electrical properties of polymer nanocomposites with carbon nanotube fillers. *Nanotechnology* 19(21):215701.
21. Bauhofer, W., Kovacs, J.Z. 2009. A review and analysis of electrical percolation in carbon nanotube polymer composites. *Composites Science and Technology* 69(10):1486–1498.
22. Oliva-Aviles, A.I., Aviles, F., Sosa, V. 2011. Electrical and piezoresistive properties of multi-walled carbon nanotube/polymer composite films aligned by an electric field. *Carbon* 49(9):2989–2997.

23. Pereira, L.F.C., Ferreira, M.S. 2011. Electronic transport on carbon nanotube networks: A multiscale computational approach. *Nano Communication Networks* 2(1):25–28.
24. Snow, E.S., Campbell, P.M., Ancona, M.G., Novak, J.P. 2005. High-mobility carbon-nanotube thin-film transistors on a polymeric substrate. *Applied Physics Letters* 86:033105.
25. Ha, M., Xia, Y., Green, A.A., Zhang, W., Renn, M.J., Kim, Hersam, M.C., Frisbie, C.D. 2010. Printed, sub-3V digital circuits on plastic from aqueous carbon nanotube inks. *ACS Nano* 4(8):4388–4395.
26. Chen, P., Fu, Y., Aminirad, R., Wang, C., Zhang, J., Wang, K., Galatsis, K., Zhou, C. 2011. Fully printed separated carbon nanotube thin film transistor circuits and its application in organic light emitting diode control. *Nano Letters* 11(12):5301–5308.
27. Meitl, M.A., Zhou, Y.X., Gaur, A., Jeon, S., Usrey, M.L., Strano, M.S., Rogers, J.A. 2004. Solution casting and transfer printing single-walled carbon nanotube films. *Nano Letters* 4(9):1643–1647.
28. Engel, M., Small, J.P., Steiner, M., Freitag, M., Green, A.A., Hersam, M.C., Avouris, P. 2008. Thin film nanotube transistors based on self-assembled, aligned, semiconducting carbon nanotube arrays. *ACS Nano* 2(12):2445–2452.
29. Meitl, M.A., Zhou, Y.X., Gaur, A., Jeon, S., Usrey, M.L., Strano, M.S., Rogers, J.A. 2004. Solution casting and transfer printing single-walled carbon nanotube films. *Nano Letters* 4(9):1643–1647.
30. LeMieux, M.C., Roberts, M., Barman, S., Jin, Y.W., Kim, J.M., Bao, Z. 2008. Self-sorted, aligned nanotube networks for thin-film transistors. *Science* 321(5885):101–104.
31. Vosgueritchain, M., LeMieux, M.C., Dodge, D., Bao, Z. 2010. Effect of surface chemistry on electronic properties of carbon nanotube network thin film transistors. *ACS Nano* 4(10):6137–6145.
32. Jung, M., Kim, J., Noh, J., Lim, N., Lim, C., Lee, G., Kim, J., Kang, H., Jung, K., Leonard, A.D., Tour, J.M., Cho, G. 2010. All-printed and roll-to-roll-printable 13.56-MHz-operated 1-bit RF tag on plastic foils. *IEEE Transactions on Electron Devices* 57(3):571–580.
33. Xiang, L., Zhang, H., Dong, G., Zhong, D., Han, J., Liang, X., Zhang, Z., Peng, L.M., Hu, Y. 2018. Low-power carbon nanotube-based integrated circuits that can be transferred to biological surfaces. *Nature Electronics* 1:237–245.
34. Zhao, Y., Li, Q., Xiao, X., Li, G., Jin, Y., Jiang, K., Wang, J., Fan, S. 2016. Three-dimensional flexible complementary metal–oxide–semiconductor logic circuits based on two-layer stacks of single-walled carbon nanotube networks. *ACS Nano* 10(2):2193–2202.
35. Wang, C., Hwang, D., Yu, Z., Takei, K., Park, J., Chen, T., Ma, B., Javey, A. 2013. User-interactive electronic skin for instantaneous pressure visualization. *Nature Materials* 12:899–904.
36. Sekiguchi, A., Tanaka, F., Saito, T., Kuwahara, Y., Sakurai, S., Futaba, D.N., Yamada, T., Hata, K. 2015. Robust and soft elastomeric electronics tolerant to our daily lives. *Nano Letters* 15(9):5716–5723.
37. Kim, J., Shim, H.J., Yang, J., Choi, M.K., Kim, D.C., Kim, J., Hyeon, T., Kim, D.H. 2017. Ultrathin quantum dot display integrated with wearable electronics. *Advanced Materials* 29(38):1700217.
38. Choi, M.K., Yang, J., Kim, D.C., Dai, Z., Kim, J., Seung, H., Kale, V.S., Jung, S.J., Park, C.R., Lu, N., Hyeon, T., Kim, D.H. 2018. Extremely vivid, highly transparent, and ultrathin quantum dot light-emitting diodes. *Advanced Materials* 30(1):1703279.
39. Chortos, A., Koleilat, G.I., Pfattner, R., Kong, D., Lin, P., Nur, R., Lei, T., Wang, H., Liu, N., Lai, Y.C., Kim, M.G., Chung, J.W., Lee, S., Bao, Z. 2016. Mechanically durable and

highly stretchable transistors employing carbon nanotube semiconductor and electrodes. *Advanced Materials* 28(22):4441–4448.

40. Kang, I., Schulz, M.J., Kim, J.H., Shanov, V., Shi, D. 2006. A carbon nanotube strain sensor for structural health monitoring. *Smart Materials and Structures* 15(3):737–748.
41. Yu, M.F., Lourie, O., Dyer, M.J., Moloni, K., Kelly, T.F., Ruoff, R.S. 2000. Strength and breaking mechanism of multiwalled carbon nanotubes under tensile load. *Science* 287(5453): 637–640.
42. Bauhofer, W., Kovacs, J.Z. 2009. A review and analysis of electrical percolation in carbon nanotube polymer composites. *Composites Science and Technology* 69(10):1486–1498.
43. Hu, N., Karube, Y., Yan, C., Masuda, Z., Fukunaga, H. 2008. Tunneling effect in polymer/carbon nanotube nanocomposite strain sensor. *Acta Materialia* 56(13):2929–2936.
44. Chen, J.K., Gui, X.E., Wang, Z.W., Li, Z., Xiang, R., Wang, K.L. 2012. Super low thermal conductivity 3D carbon nano tube network for thermoelectric applications. *ACS Applied Materials & Interfaces* 4(1):81–86.
45. Kim, Y.S., Kim, D., Martin, K.J., Yu. C., Grunlan, J.C. 2010. Influence of stabilizer concentration on transport behavior and thermopower of CNT-filled latex-based composites. *Macromolecular Materials and Engineering* 295(5):431–436.
46. Choi, K., Yu, C. 2011. Influence of nanoparticles on thermoelectric properties of organic composite. International Proceedings of 18th International Conference on Composite Materials.
47. Kim, G., Pipe, K.P. 2012. Thermoelectric model to characterize carrier transport in organic semiconductors. *Physical Review B: Condensed Matter* 86(8):085208.
48. Lisunova, M., Mamunya, Y.P., Lebovka, N., Melezhyk, A. 2007. Percolation behaviour of ultrahigh molecular weight polyethylene/multi-walled carbon nanotubes composites. *European Polymer Journal* 43(3):949–958.
49. Logakis, E., Pissis, P., Pospiech, D., Korwitz, A., Krause, B., Reuter, U. 2010. Low electrical percolation threshold in poly (ethylene terephthalate)/multi-walled carbon nanotube nanocomposites. *European Polymer Journal* 46(5):928–936.
50. Kim, Y.J., Shin, T.S., Do Choi, H., Kwon, J.H., Chung, Y.C., Yoon, H.G. 2005. Electrical conductivity of chemically modified multiwalled carbon nanotube/epoxy composites. *Carbon* 43(1):23–30.
51. Zare, Y., Rhee, K.Y., Park, S.J. 2019. Modeling the roles of carbon nanotubes and interphase dimensions in the conductivity of nanocomposites. *Results in Physics* 15:102562.
52. Zare, Y., Rhee, K.Y. 2019. Simplification and development of McLachlan model for electrical conductivity of polymer carbon nanotubes nanocomposites assuming the networking of interphase regions. *Composites Part B: Engineering* 156:64–71.

8 Optical Sensor-Based Hydrogen Gas Detection

A Present View

Archita Lenka, Bandita Panda, Chinmaya Kumar Sahu, Narayan Panda, and Sandip Kumar Dash

CONTENTS

8.1 INTRODUCTION

In the present era, humans are facing a major environmental crisis due to the reliance on fossil-based fuels, which is worsening with time. A possible solution is to use renewable energies like solar, wind, or hydro power. Apart from these, hydrogen (H_2) is another promising substitute by virtue of its abundance and non-polluting properties [1]. However, a well-defined and systematically organized storage and dissemination system is essential to successfully realize the energy source [2]. H_2 has long been used as a source of energy in different health, chemical, and manufacturing sectors [3]. Also, it is released as a byproduct from many physiochemical or biological processes [4]. Occasional cases of H_2 production have also been reported from the waste tanks of nuclear plants, which is a matter of serious concern, as it could lead to nuclear explosion [5]. If free H_2 reacts to the O_2, halogens, or strong oxidants in the atmosphere in the presence of a catalyst, this can result in explosion [6] and at a higher atmospheric concentration, can disbar the free supply of O_2 to living organisms [7–9] causing blue skin, headache, nausea, vomiting, fever, and/or asphyxia [10]. In addition, if liquid H_2 is accidentally exposed to living organisms, it can behave as a cryogen to cause frosting [11]. Therefore, the detection and measurement of H_2 gas from different sources has been a matter of extreme priority in order to avoid any unwanted consequences. H_2 is a colorless and odorless gas with a relative density of ~0.07 [12]. Being the lightest element, it can easily leak out from any source [13]. Further, since the ignition energy of the gas is very low (~0.02 mJ), it can explode on heating at high temperature [14]. Further, the explosive limit of the gas is approximately 4–75% (v/v), and, therefore, it is highly inflammable, making it extremely hazardous and unsafe [15].

These uncommon physicochemical properties of the gas raise concern about its detection, which requires very high accuracy and sensitivity. Several researchers have tried in different ways to establish efficient detection of the gas. The conventional methods used for H_2 detection include gas chromatography (GC), GC-thermal conductivity (GC-TC), GC-mass spectrometry (GC-MS), and laser gas analysis. But recently, the research world has been enormously attracted to H_2 sensors as a substitute to the conventional methods because of their simplicity, portability, sensitivity, specificity, low cost, and reusability. The sensors used for this are mainly semiconductor [16, 17], electrical [18, 19], or optical based [20]. Above all, optical sensors hold a special position because of their sensitivity, specificity, and reproducibility [21]. Optical sensors detect the direct interaction between light and H_2 or the H_2-induced topological change of the transducer. Fourier transform infrared spectroscopy (FTIR) or Raman spectroscopy can directly detect H_2, both qualitatively and quantitatively, whereas surface plasmon resonance (SPR), interferometry, or fiber grating analysis (FGA) are used to detect the surface-H interaction.

This chapter explores different types of methods used for the detection of H_2 from different sources, focusing especially on the types of optical sensors that exist, along with their advantages and disadvantages.

8.2 GC-BASED DETECTION

The conventional methods used for detection of H_2 are quite time-consuming, cumbersome, and involve instruments which are costly and technical to handle [22]. Therefore, researchers have sought a simple, portable, and economical yet sensitive, specific, and reproducible method by using recently advanced devices like microarrays, nanochips, or nanosensors. One of the most widely used conventional techniques for the sensitive and efficient detection of H_2 is GC. A GC basically separates the components of a gaseous mixture on the basis of their interaction with inert gases and the difference in columnar retention time [23]. This method is best suited for qualitative and quantitative detection of gases which vaporize easily on heating without decomposing (Figure 8.1) [24]. Falony et al. (2009) analyzed the growth kinetics of five different butyrate-producing bacterial strains by studying their rate kinetics for breakdown of the carbohydrates into butyrate, CO_2, and H_2 using GC [25]. Similarly, Traore et al. (2019) cultured *Methanobrevibacter smithii* (an H_2 removing bacteria) mixed with different H_2-producing bacteria and estimated the H_2 removal efficiency of the bacteria [26]. Another group detected the release of H_2 from ammonia borane (NH_3BH_3) by attaching a mini whistle into the outlet of a GC, whose change in the frequency was detected using an accelerometer [27]. The mechanism of auxin-induced H_2 gas production in the plants leading to the origin of lateral roots through the regulation of NO was studied by Cao et al. (2017) in tomato seedlings using GC [28]. In the same year, Fan et al. (2017) devised a solid oxide fuel cell detector and used a double logarithmic model to improve the specificity of H_2 detection [29]. Similarly, Ganzha et al. (2018) put forward an approach to strain the impurities (N_2 and O_2) present in H_2 and deuterium using GC in an attempt to obtain pure H_2 [30] Similarly, Abrahim et al. (2020) used high temperature Cr-Ag reactor-based GC to estimate the isotopes of H_2 released from the breakdown of mono or disaccharides from food or beverages [31]. This GC-based detection is sensitive and specific, but also laborious and time-consuming [32–34].

The researchers tried for TC detectors (TCDs) to improve the detection response, accuracy, and sensitivity of the detection [35]. A TCD consists of four thermistors connected via a Wheatstone bridge circuit [36]; when heated at a constant current, heat is released into the enclosed chamber, which is directly related to the TC of the surrounding gas [37]. The reference chamber carries the inert gas, whereas H_2 flows through the sample chamber. The relative difference of the electrical resistance of the

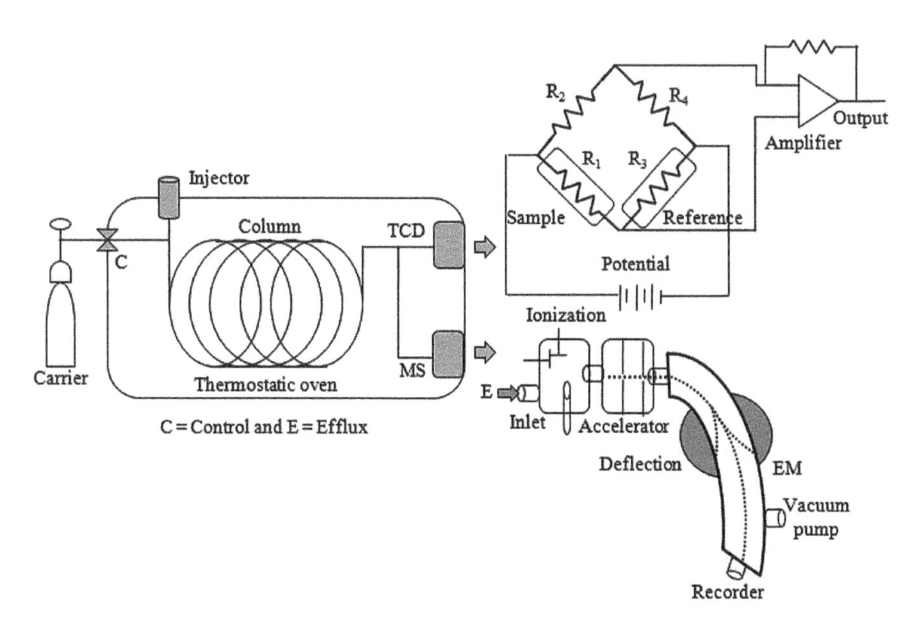

FIGURE 8.1 Schematic presentation of GC-TC and GC-MS for the detection of H_2.

thermistors is detected both quantitatively and qualitatively [36, 38]. Hubert et al. (2014) used GC-TC for volumetric analysis of H_2 [39].

H_2, being light, volatile, flammable, and thermo-conductive, is hard to detect and quantify, but GC-MS provides an excellent solution. The technique is highly preferred for the isotopic separation of H_2. The gas effluxes of a GC are ionized and accelerated through a vacuum chamber fitted with electromagnets for detection. Similarly, Ellefson et al. (1981) used quadrupole mass spectrometry (MS) to detect and quantify the H_2 from a mixture of its other isotopes [40]. Later on, Genty and Schott, (1970) focused on differentiation of the six different types of isotopes of H_2 by using a ferric-hydroxide-coated, alumina-filled GC-MS [41]. Similarly, another group separated the six isotopes of H_2 using a liquid N_2 cooling column GC-MS [42, 43]. Gehre et al. (2015) used an elemental analyzer/high-temperature conversion instrument for isotopic detection of H_2 from N_2 containing compounds [43], and Nair et al. (2015) did the same [44]. In the next year, Gehre et al. (2016) introduced a Cr-filled section in the region of 200–1,200°C into the instrument to enable the detection from halogen- and sulphur-containing compounds too [45]. Seeing the need for a hydrogen-deuterium exchange (HDX) study to understand the structure and composition of a small organic molecule, Eysseric et al. (2016) used a GC-MS-based post-column HDX to study different organic molecules [46].

Takai et al. (2002) visualized the concentration and state of the H_2 and its desorption from ferrite, graphite, and pearlite using secondary ion mass spectrometry (SIMS) [47]. Their study found that heating at 100, 300, and 200°C caused desorption of the H_2 from ferrite, pearlite, and the interfaces between the ferrite and graphite, respectively. But it has been seen that during the study of hydrogenous content of a metal using GC-MS, the H_2 present in the moisture, hydrocarbons, or other organic materials present on the sample or MS chamber itself can mislead H_2 detection. Therefore, Awane et al. (2011) used an Si sputtering method to measure and reduce these background contaminants [48]. Recently, Kaminsky et al. (2020) detected the volume of H_2 present in Brazilian diamonds using SIMS [49]. Recently, Camargo et al. (2020) employed GC-MS to assess limonene in batch reactors, which in turn quantified the H_2 production at a temperature of 30 °C [50]. So too, Tang et al. (2020) studied the H_2 released from different organohalides through GC-MS using $^{13}C_6Cl_6$ and $^{13}C_6Cl_6$ and found the highest amount of release at 45 eV EI energy and 0.9 mA [51].

8.3 SEMICONDUCTOR-BASED SENSORS

Although numerous methods have been proposed for H_2 detection, very few have been commercialized. Because of this, the demand for a portable, sensitive, specific, and economical detection is still high. Recently, H_2 sensors have emerged as one of the most suitable substitutes for earlier methods, and optical sensors are of particular interest due to their sensitivity, specificity, low cost, and reproducibility. The performance of a sensor is determined by its sensitivity, specificity, limit of detection (LOD), and stability. Specificity of a sensor is vital to obtain minimum false positive or negative results [52], whereas sensitivity and LOD defines the efficacy of detection. Simultaneously, stability of a sensor is among the most desirable qualities because it reduces the cost of the detection significantly [53]. Moreover, an ideal H_2 sensor must be operationally simple and robust [54] and must not be affected by any contaminants. The most important criteria for commercialization of a sensor in developing countries like India is its economic value [55] so that it can be used in devices for the general public. The sensors used for the detection of H_2 are mainly semiconductor, electrical, semiconductor, electrochemical, or optical based.

Semiconductor-based H_2 sensors are known for their simplicity, portability, sensitivity, specificity, and economic value [56]. These sensors are fabricated by different combinations of metals, oxides, and insulators, and they are categorized according to their detectable physicochemical properties. Because of their good H_2 adsorption efficiencies, metals [57] or their alloys [57–59] are preferred for fabricating sensors. Chemiresistors measure the H_2-induced modulation of the resistance or conductance at their transducers [60]. The nanoparticles (NPs) of oxides like SnO_2 [61], ZnO [62], TiO_2 [63] or WO_3 [64] are often fabricated onto the sensing films and are sometimes doped with metals to improve the overall sensitivity or performance [65, 66]. Some reports suggest the use of nanocomposites on these sensing layers to increase adsorption [16], whereas a few others advocate the use of carbon nanotubes to retain high surface area and stability [67, 68].

Metal and semiconductor paired Schottky diode-based H_2 sensors involve measurement of the output current at a constant temperature and voltage by means of their interfacial Schottky barrier height [69]. The H_2 molecules adsorb and then penetrate into the interface, where they are polarized by the external voltage leading to a decrease in the barrier height [70, 71]. Further, to improve sensitivity, the Schottky barrier height is increased by introducing an extra layer of semiconductor or oxide into the interface [72]. Another such semiconductor-based H_2 sensor is the field effect transistor (FET) sensors [73]. Just like the Skhottky diodes, in FET sensors, H_2 enters the interface, and the change in the electrochemical properties of the floating gate [74] in the form of voltage or current is detected by keeping other parameters between the source and drain constant [75–77]. However, in capacitor H_2 sensors the H_2 at the interface is detected as the change in capacitance [78]. Although semiconductor-based sensors consume low levels of energy [79], they sometimes discharge sparks and cause explosions [80].

8.4 ELECTRICAL SENSORS

Electrical H_2 sensors generally convert the H_2-induced physicochemical changes on the surface of the transducer into an electrical signal. These sensors can be of an acoustoelectric, a thermoelectric, or a mechano-electric type. Acoustoelectric sensors are mainly constructed by coating a layer of sensing material on a piezoelectric material (PEM) or PEM-metal oxide composite substrate [81]. Further, interdigited electrodes at the interface of these sensors convert the surface acoustic waves (SAW) into an electrical signal [82]. The H_2 at the interface changes the electrical properties of the sensing layer and hence shifts the central frequency of the SAW on the surface through an acoustoelectric effect [83, 84], which is detected. Thermoelectric sensors basically measure the temperature change of the transducer by the heat energy release from catalytic oxidation of H_2 [83, 85], whereas in the mechano-electric sensors, the change in the vibrational frequency, bending, or deflection of the microcantilevers due to the adsorption of H_2 molecules is converted into an electrical signal [19]. Although these electrical sensors are quick in their response and reproducible, LOD is always an issue. Also the sensitivity or specificity of detection are influenced by environmental conditions, therefore, sometimes these are less attractive.

8.5 TC SENSORS

H_2 is known to exhibit high TC, and therefore several TC-based sensors have been studied for the detection of H_2. A micromachined multi-sensor was reported for the volumetric estimation of H_2 by aggregating the temperature, humidity, and TC together [86]. Another multi-sensor was put forward by Leonardi et al. (2018), with amalgamation of resistance and TC for the detection of H_2 [87]. In the same year, Romanelli et al. (2018) used neutron transmission and TC together to evaluate the quantity of parahydrogen [88]. Tardy et al. (2004) were the first to report an H_2 sensor operated at a low-temperature (37 °C) [89], and to simplify the detection of H_2 even at higher temperatures, Li et al. (2020) used a short-hot-wire process [90]. Apart from this, TC sensors can also be used to detect H_2 from a gaseous mixture; for example, Simon and Arndt (2002) detected H_2 (up to 2%) from automobile exhausts using a TC sensor [91]. More recently, a TC sensor was reported to record up to 2,000 ppm of H_2 from a gas mixture [37].

8.6 ELECTROCHEMICAL SENSORS

Electrochemical sensors have always been preferred in the design of simple but sensitive and specific sensors for detection of any analyte. Detection of H_2 requires high precision and specificity, and electrochemical sensors have potential to stand up to the expectation of researchers to fulfill this demand. A basic electrochemical sensor can be devised using three electrodes (reference, working, and counter electrodes; RE, WE, and CE) in which the required analyte is detected on the working electrode in the form of a current or voltage output between the working and reference electrodes through applying a constant voltage or current between reference and counter electrodes. Adsorption of the H_2 onto the surface of the working electrode is converted into an electrical signal (Figure 8.2).

A Pd-nanoribbon array was fabricated on a polyethylene terephthalate substrate for an electrochemical H_2 sensor was developed at room temperature (RT) using a direct transfer method [92]. An electrochemical H_2 sensor was used by Zhao et al. (2018) to detect H_2 from bone marrow cells [93]. In another experiment, the working electrode was fabricated by electrophoretic deposition of PdO-reduced graphene oxide NPs onto an indium tin oxide-coated glass for the design of an H_2 sensor. The sensor could detect 10–80% of H_2 at a sensitivity of 0.462 μA/% [94]. Holstein et al. (2019) designed and fabricated an H_2 sensor prototype [95]. A composite of polyaniline-zinc-zeolitic benzimidazolate framework was fabricated for electrochemical H_2 sensor development [96].

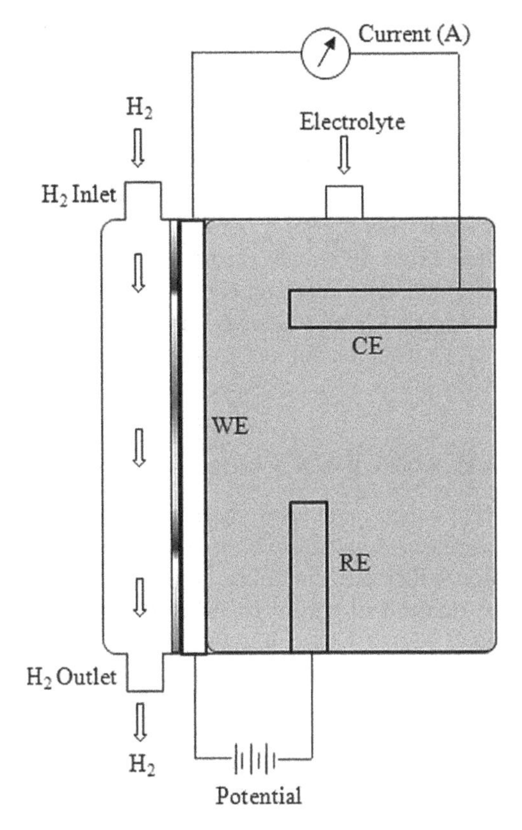

FIGURE 8.2 Schematic presentation of an electrochemical H_2 sensor.

8.7 OPTICAL SENSORS

Among the different H_2 sensors available today, optical sensors hold a special position because of their sensitivity, specificity, and reproducibility. Optical sensors can directly detect H_2 from a source using FTIR or Raman spectrometry, or they can detect H_2-induced optical changes of transducers through intensity variation, SPR, interferometry, or FGA.

8.7.1 FTIR-Based Sensors

FTIR is one of the simplest methods for detecting H_2 directly from different metallurgical sources, and it is often used as a confirmatory test against results obtained with other methods of detection. Khisina et al. (2001) used FTIR along with other techniques like transmission and analytical electron microscopy for the detection of H_2 from olivine crystals [97]. In subsequent years, Koga et al. (2003) detected the H_2 content of different natural minerals using FTIR and utilized the values to calibrate SIMS-based detection [98]. The group was able to lower the LOD up to 2–4 ppm.

Aubaud et al. (2007) reported discrepancies in the results of FTIR and SIMS-based detection of H_2 from 8 basaltic glasses and 11 rhyolitic glasses, 17 olivines, 9 orthopyroxenes, 8 clinopyroxenes, and 8 garnet samples; this may be due to the moisture content of the samples [99].

The work by Pan et al. (2011) emphasized adsorption of H_2 in Ga_2O_3 followed by dissociation of the substrate into OH and GaH, which was detected using FTIR and periodic density functional theory [100]. Panayotov et al. (2015) used a combination of FTIR of the donated electrons, co-adsorbed CO, and HDX to provide a detailed insight into the Rh/TiO_2-mediated reversible breakdown of H_2 [101]. The release of H_2 through elution due to the chemical conversion of $NaAlH_4$ to Na_3AlH_6 and then into NaH was analyzed by a combination of spectral and gravimetrical measurement using FTIR and an X-ray diffractometer, respectively [102].

8.7.2 Raman Spectroscopy-Based Sensors

Raman spectroscopy, although quite costly because of its sophisticated instrumentation, is nevertheless sensitive and specific in H_2 detection. In this method, the H_2 molecules in proportion to their concentration absorb a part of the incident laser beam, causing a shift in the transmitted light wavelength (Figure 8.3) detected through a charge coupled device (CCD). Adamopoulos et al. (2004) used visible Raman spectroscopy for the detection of H_2 from hydrogenated amorphous carbon by elastic recoil detection analysis-Rutherford back scattering [103]. Sturm et al. (2009) used Raman scattering to differentiate pure H_2 (up to 0.1%) from its other isotopes with high accuracy [104]. Numata et al. (2013) devised a Raman-based gas cell for the qualitative and quantitative analysis of H_2 from fermentation [105]. Recently, Gao et al. (2019) was able to separate different components of natural gas including H_2 with the help of automatic decomposition algorithm of Raman spectroscopy [106]. Li et al. (2015) reported that Raman can also be used for quantitative and qualitative analysis of molten H_2 [107].

One of the major disadvantages of Raman spectroscopy is its weak signal strength, and photoacoustic stimulation using a laser beam can increase the spectral dispersion and hence the sensitivity [108]. Tedder et al. (2010) detected H_2 up to 0.03 M using anti-Stokes Raman and enhanced the sensitivity using duel-pump measurements [109, 110]. Two years later, Spencer et al. (2012) detected 40 ppm and 4.6 ppm H_2 by using pulsed Nd:YAG and dye lasers, respectively, at 1 atm N_2 [111]. Simultaneously, they could even differentiate between the ortho and para form of H_2. In an alternative approach, Wen et al. (2019) tried with a four-mirror multiple pass instead of the conventional two-mirror Raman setup to increase sensitivity of detection, along with high signal-to-noise ratio [112]. In the next year, they constructed a 20-pass cavity by using three-mirror setup to decrease the LOD and increase the sensitivity further [113].

Hanf et al. (2015) and Knebl et al. (2018) used fiber-enhanced Raman spectral analysis for the detection of H_2 from exhaled gases (up to 4.7 ppm) and gaseous mixtures, respectively [114, 115]. Qi et al. (2019) devised a nanofiber sensor based on Raman spectroscopy to detect H_2 with 100% accuracy and a sensing time < 10 seconds [116].

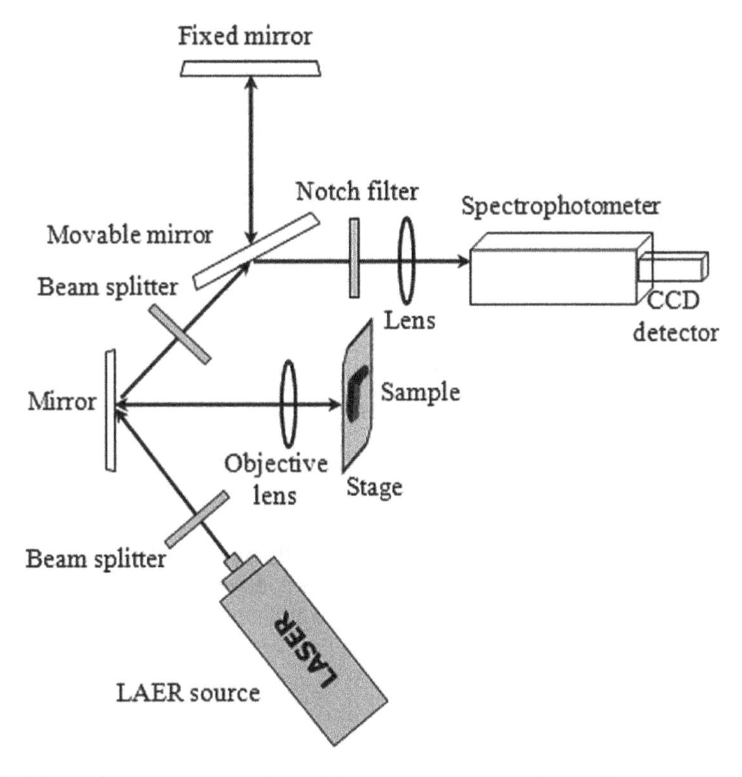

FIGURE 8.3 Schematic presentation of the principle of detection of H_2 using conventional Raman spectrometry.

8.7.3 SPR-Based Sensors

Although, FTIR and Raman are fast, pollution free, and non-destructive, they prove inefficient for detecting diatomic molecules like H_2 [117]. Therefore, Chadwik and Gal (1993) for the first time reported a Pd-coated SPR sensor for the detection of H_2 [118], and the next year the Pd was substituted by a Pd/Ni alloy layer to improve sensitivity [119]. Photonic crystal fiber, due to its light contrasting capability, was used to develop an SPR sensor for the separation of H_2 from a mixture containing methane and H_2 [120]. Cavalcanti et al. (2009) deposited a film of Pd onto a glass prism to develop an SPR sensor [121], but later on, it was realized that due to their large size, prism-based sensors are difficult to use in remote sensing or other mini apparatus [122]. Therefore, Watkins and Borensztein, (2018) reported a Pd nanofilm-deposited glass surface for fabricating a localized SPR [123]. Recently Shafieyan et al. (2019) and Wang et al. (2020) reported two more sensors of the same type [124, 125].

8.7.4 Optical Fiber Sensors

Optical fiber sensors show promise over other optical sensors for the detection of H_2 because of their size, sensitivity, and stability [126]. In an optical fiber sensor, the fiber itself can act as an intrinsic sensor [127, 128] or may be coated with a Pd/WO_3 film, either at their end or on their lateral wall, to act as the sensing film. The H_2-induced optical change of the film is detected through Raman, SPR, interferometry, or FGA. On close contact with Pd, the H_2 molecules dissociate into atoms and diffuse through the layer, causing its expansion and a decrease in the dielectric constant [129] (Equations 8.1 and 8.2). These physicochemical changes lead to change in optical properties [130].

$$H_2 \xrightarrow{Pd} 2H \tag{8.1}$$

$$\alpha Pd + H \longrightarrow \beta PdH \tag{8.2}$$

Mechanical stability of Pd-film-based optical fiber sensors is always a matter of concern, as the metallic expansion may lead to damage, delamination, or cracking [131]. Therefore, slowly, the metal was substituted by metallic-doped forms. But this has put a big question mark on their sensitivity and specificity, and researchers started to prefer WO_3. The film of this oxide can undergo visible color change when exposed to H_2 [132, 133]. On reacting to H_2, the oxide layer produces H_2O, and in the presence of the atmospheric O_2, reverses back to its original oxide form for further use (Equations 8.3 and 8.4). In spite this, WO_3 cross-reacts with other gases during H_2 detection; therefore, it can be resolved through metallic-doped oxide films [134, 135]. However, this metallic-doped oxide layer reversibly reacts exothermically with H_2, creating a safety concern.

$$WO_3 + H_2 \longrightarrow WO_2 + H_2O \tag{8.3}$$

$$WO_2 + 1/2O_2 \longrightarrow WO_3 \tag{8.4}$$

8.7.4.1 Optical Fiber Sensors For Intensity Change

Optical fibers sense the difference between the intensity of the incident and re-emitted light induced by the H_2. Evanescent field (EF) sensors are quite simple optical sensors, having a sensing film on a tapered section, polished on one side and with etched regions on both sides, and cladded in the remaining sections (Figure 8.4 (a)). The effective refractive index (μ_{eff}) of the sensing film will change on exposure to H_2, and so will the absorbance intensity (I) of the re-emitted light from the fiber, following Beer-Lambert's equation (Equation 8.5) [136], where, ε, c, and L are the molar absorption coefficients, concentration of the material used for sensing film, and path length, respectively. The change in intensity is directly proportional to the change in the μ_{eff} of the sensing film which is proportional to the concentration of the H_2 adsorbed onto the film.

$$\varepsilon cL = -\log_{10} I / I_0 \tag{8.5}$$

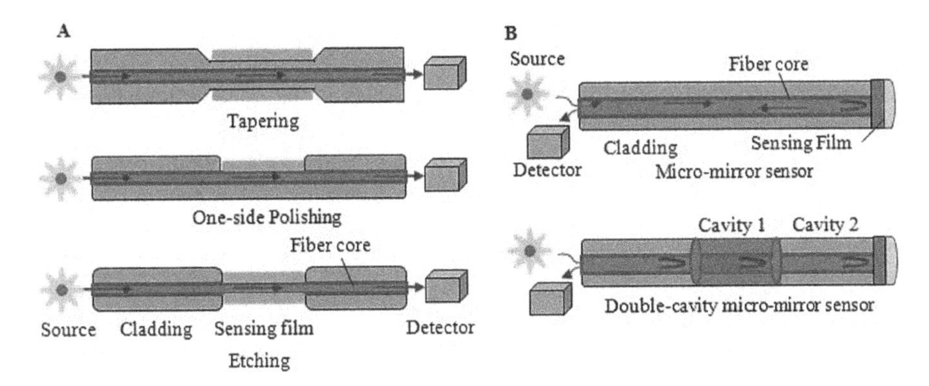

FIGURE 8.4 Schematic presentation of different types of optical fiber setup in (a) the EF and micro-mirror H_2 sensors and (b) the micro-mirror-based optical fiber sensors.

Ever since the first Pd-film-based EF H_2 sensor with sensitivity of 0.2–0.6% and response time of 30–20 seconds at RT [137], several other studies have been carried out with different modifications to improve the sensitivity, specificity, or response time. Sekimoto et al. (2000) used Pt-doped WO_3 film and found better sensitivity [138]. Villatoror et al. (2001) proposed a tapered optical fiber-based EF sensor with sensitivity up to 1.8–10% and found that the sensitivity is inversely proportional to the diameter of the tapering [139]. Subsequently, a one-side polished optical fiber sensor could detect H_2 (up to 4%). A study carried out by Yang et al. (2010) on Pd and Pd/WO_3 composite-coated single- and multi-mode fibers suggested that Pd-coated fibers do not vary in sensitivity and response linearity remains the same for both cases of Pd-coated; however, Pd/WO_3 composite-coated fibers showed better sensitivity and response linearity for multimode and single-mode fibers, respectively [140]. Very recently, Pd NP was produced from $PdCl_2$ through the drop-casting technique and coated onto a tapered optical fiber for fabrication of an H_2 sensor with a detection range of 0.125–1% at RT [125, 141].

Micro-mirror sensors are mainly based on the optical fibers cladded throughout their length and a sensing film coated at one of the ends or a cleaved end (Figure 8.4 (b)). The light entering from one end undergoes Fresnel reflection before re-emitting out, but H_2 can modulate the μ of the sensing layer and hence the pitch of the light. A simple micro-mirror-based Pd-film H_2 sensor was reported by Butler et al. (1991), and Liu et al. (2012) used Pd/Y composite film to improve the signal-to-noise ratio [142, 143]. However, they had noticed a stability issue and improved it through stimulated annealing [144]. Bevenot et al. (2000) used a Pd micro-mirror-based sensor to detect leakage of H_2 from the cryotechnique engine of the European rocket ARIANE V [145]. Other groups tried Pd-capped Mg-Ti [146] and Pt/WO_3 [107] to improve the sensitivity and response linearity. Optical loss is one of the major drawbacks of micro-mirror sensors, and Park et al. (2011) and Tang et al. (2015) introduced dual-cavity-based sensors to overcome this problem [147, 148]. Although these intensity-based optical fiber sensors are simple and economical, they suffer from sensitivity and specificity issues.

8.7.4.2 SPR Optical Fiber Sensors

SPR-based optical fiber sensors are known to be sensitive, safe, and reliable for the detection of H_2 [158], and these optical fibers are coated with one to three layers of metal, metal oxides, or nanocomposites to enhance their sensitivity several times over (Figure 8.5). Bhatia and Gupta (2012) reported an SPR optical fiber sensor constructed with a sensing film from three layers, one each of Ag, Si, and Pd [122]. In another sensor, In_2O_3 and SnO_2 were coated at 70:30 [159]. In a similar fashion, Perrotton et al. (2013) used multimode fiber coated by Au, SiO_2, and Pd [160], whereas Hosoki et al. (2013) used Au, Ta_2O_5, and Pd to detect H_2 up to 4% [129]. Recently, extending their previous work, Hosoki et al. (2019) reported an H_2 sensor coated in multiple layers of Au/Ta_2O_5/Pd/Pt [157]. Similarly, many other groups, like Wang et al. (2013), Tabasum et al. (2015), and Takahashi et al. (2016), have developed optical fiber SPR sensors by coating with Pt/WO_3, $ZnO_{(1-x)}Pd_x$, Au/Ta_2O_5/Pd/Au nanocomposites, respectively [161–163]. Photonic crystal fiber, known for its light contrasting capability, was utilized by Liu et al. (2018) in an SPR sensor for H_2 detection [120]. Different reported fiber optic-based SPR sensors are listed in Table 8.1.

8.7.4.3 Interferometry Optical Fiber Sensors

Interferometer-based H_2 sensors are simple, flexible, and sensitive [164]. The first interferometry optical fibre H_2 sensor was developed by Butler in 1994 [165]. In this type of sensor, ≥ 2 laser beams of the same wavelength but different phase will be put through different core fibres of the same diameter and length under a single cladding to interfere before emitting out. The H_2-induced change in the sensing film may alter the phase difference (φ) and hence the interference pattern of emitted light. The φ is expressed in Equation 8.6 [131]. Where, λ and L are wavelength and path length, respectively. There are different optical fibre-based interferometer setups. In a Mach-Zehnder interferometry (MZI) setup, the light from a single source after traveling through different core fibers meets at a point and interferes, which can be changed by coating the fibre with a sensing film [166, 167].

$$\varphi = 2\pi / \lambda(\mu_{eff} L) \tag{8.6}$$

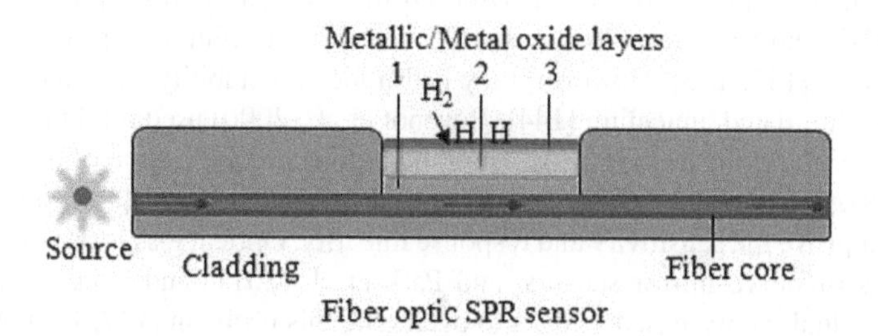

FIGURE 8.5 Schematic presentation of a fiber optic SPR H_2 sensor.

TABLE 8.1
Studies on optical fiber-based SPR sensors

Serial No.	Sensing film	Thickness of film (nm)	Method used	Concentration of detection (%)	Response time (seconds)	Carrier gas used	Wavelength shift (nm)	Reference
01	Au, Ta_2O_5, Pd	Pd: 5 and 7 Ta_2O_5: 60 Au: 25		0–4	40	N_2	28	[149]
02	In_2O_3/SnO_2	70 90	Intensity modulation interrogation	4		N_2		[150]
03	Au/Ta_2O_5/Pd	Au: 25 Ta_2O_5: 60 Pd: 5	Light intensity based	4	15	N_2	28	[151]
04	Au/SiO_2/Pd	Au: 35 SiO_2: 180 Pd: 2.5	Wavelength modulation	0.5–4.0	15	Ar	640–680	[152]
05	Pt/WO_3	Ag: 35 SiO_2: 100 WO_3: 180 Pt: 3	Wavelength modulation	4		Ar	17.4	[153]
06	$ZnO_{(1-x)}Pd_x$	94		0–4	60	N_2	~ 80	[154]
07	Au/Ta_2O_5/Pd/Au	Au: 25 Ta_2O_5: 60 Pd: 1.4 Au: 0.6		0–4	3–6	N_2	Pd/Au: 14.4 Pd: 24.4	[155]
08	Au/Pd-WO_3		Side hole structure and polarization filtering	0–3				[156]
09	Au/ Ta_2O_5/Pd/Pt (Pt)	Pt: 1, 2, or 7 Au: 1 Ta_2O_5: 60 Au: 25		4	1–12	N_2		[157]

A classical MZI H_2 sensor with Pd coating on a cladded optical fibre was reported by Kim et al. (2010) [168]. Subsequently another MZI sensor with a sensitivity of 0.155 was reported by Wang et al. (2012) for H_2 detection using air-cavity-driven optical wave phase separation [39]. Similarly, Zhou et al. (2014) used a photonic crystal fibre coated with Pd on its lateral side and one end to develop an H_2 sensor [169]. Subsequently, Yu et al. (2016) reported a Pd-film-coated tapered-fibre MZI sensor [170]. But Pd-coated MZI sensors are known to be environmentally sensitive; therefore, different types were tried for other alternatives [171]. As a result, an MZI sensor consisting of two peanut-type segments coated with polydimethylsioxane-WO_3/SiO_2 film (5 mm) was developed for the detection of H_2 in a concentration ranging from of 0.0% to 0.901% [172].

The Faby-Perrot interferometer (FPI), just like a reflective device [173], detects the shift in the effective wavelength of the re-emitted light due to interference between light beams of differential phase. The reflective intensity is expressed in Equation 8.7 [131]. The detection can be either intrinsic or extrinsic depending on the location of the sensing film. The intrinsic FBI measures the change in the wavelength of the resultant reflectance from two reflectors separated by a cavity. Caibin et al. (2016) proposed a three-interface reflectance-based FPI made up of a fiber core end, air cavity, and a Pd-Y film-coated multimode fiber end for H_2 detection [174]. Li et al. (2018) developed a small FPI sensor containing an arm of large area and another arm of Pt-WO_3/SiO_2-coated hollow-core fiber [175] to detect H_2 (0.0–2.4 %) with sensitivity −1.04 nm/% within ~ 80 seconds. The cavity in an FPI optical fiber sensor can be generated simply by separating the two fibers. A recent report by Zhou et al. (2020) illustrated an FPI formed by placing one flat fiber with an end facet and a second one cladded with ceramic and coated with Pd to detect H_2 with 1.4 nm/% sensitivity in 3 minutes and 5 seconds [176]. Cao et al. (2020) used a Pd alloy-coated intrinsic FPI sensor array for the detection of up to 0.25% H_2 from seven different locations at a time [177].

$$I = I_1 + I_2 + 2\sqrt{I_1 I_2}\, \mathrm{Cos}[4\pi / \lambda(\mu_{eff} L)] \quad (8.7)$$

In the case of extrinsic FPI, the interfaces between the fiber and the sensing film acts as a first reflector, whereas the interface between the film and surrounding environment acts as a second one (Figure 8.6). In an experiment, a Pd (5.6 nm)/multilayered graphene (3 nm) composite was used at the end of an FPI H_2 sensor to detect up to 20 ppm H_2 in about 18 seconds [178]. Another extrinsic FPI sensor was reported recently by filling a Pt/WO_3-coated silica capillary with thermo-sensitive liquid to detect around 0–4% of H_2 [179].

Other types of interferometric optical fiber sensors are the Sagnac interferometry (SI) sensors, also known as fiber-within-fiber sensors [180]. In these sensors, a fiber is inserted into another fiber loop to separate the path length of the propagating waves followed by interference. Yang et al. (2015) developed an SI sensor by inserting Pd/Ag alloy-coated polarization-maintaining photonic crystal fiber into a fiber loop for the detection of H_2 up 1–4% [181]. An SI-based H_2 sensor was developed in which the polarization maintaining fiber was covered with a Pt-loaded WO_3/SiO_2 layer for

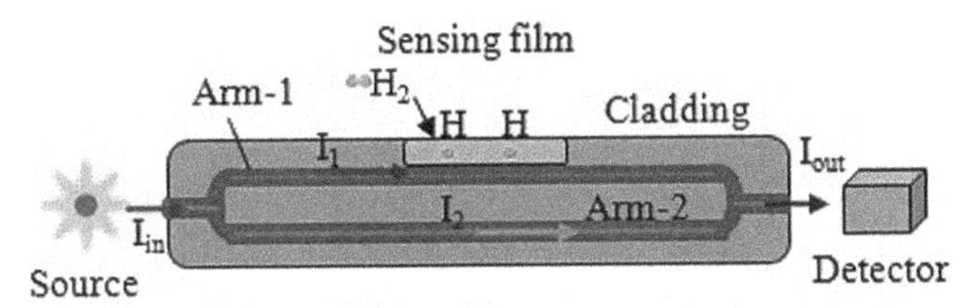

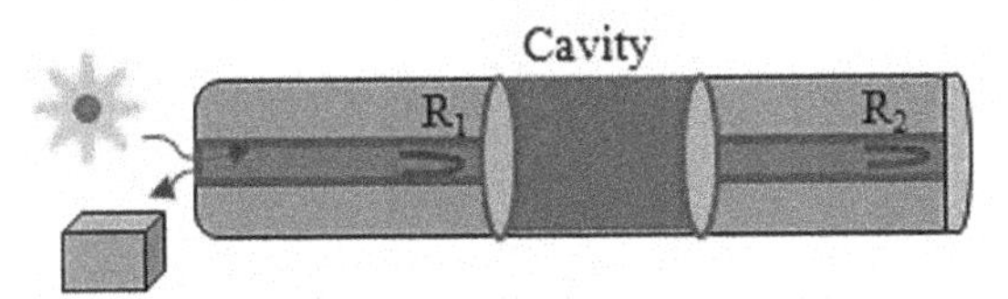

FIGURE 8.6 Schematic presentation of Mach-Zehnder and FPI interferometry optical fiber H_2 sensors.

the detection of H_2 with a sensitivity of −7.877 nm/% for 0–1% H_2 concentration [182, 183]. Another temperature-sensitive SI sensor was developed using Pt-loaded WO_3/SiO_2 film [184]. Different reported interferometry-based optical fiber sensors are listed in Table 8.2.

8.7.4.4 Fiber Grating Optical Sensors

In this type of sensor, the core of the fiber is grated at a regular interval of < 1 or 10–100 μm for fiber Bragg grating (FBG) or long periodic grating (LPG) analysis, respectively [200]. On interaction with the H_2 the sensing film either expands (pd) [201] or liberates heat (WO_3) [202], which causes a change in the periods (Λ) or the μ_{eff} of the grading. This ultimately shifts the Bragg wavelength (λ_{FBG}) of the emitted light proportional to the concentration of the H_2 (Equation 8.8) [203]. Similar to the evanescent field sensors, FBG sensors are fabricated using etched, tapered, or one-side polished fibers (Figure 8.7).

$$\lambda_{FBG} = 2\mu_{eff} \wedge \tag{8.8}$$

A Pd-deposited FBG H_2 sensor was developed for separation of H_2 from a mixture containing N_2 [204]. Another FBG sensor was developed by magnetron-sputtering-based deposition of a Pd film, followed by heating via an infrared power laser light to sense H_2 at even < 0.5% within 10 seconds [205]. For the detection of H_2 from the transformer oil, an FBG sensor with comparatively high temperature resistance was developed by loading with polyimide and coating with a Pd film [206]. Two years later, two different FBG sensors were developed, one coated with only with Pd and the other with a Pd/Cr mixture (58:42). Rhe results showed that the simple Pd was more sensitive than the mixture for detecting dissolved H_2 from the transformer

TABLE 8.2
Interferometric optical fibre-based sensors for detection of H_2

Serial No.	Type of sensor	Sensing film	Sensitivity	Concentration of H_2 detected	Response time	Reference
01	MZI	Pd	0.1 nm/%	4%	10 minutes	[185]
02		Pd	0.25 nm/%	0–5%	15 seconds	[186]
03		Pt-doped WO_3	~ 660 pm/%	0–4%	~ 120 seconds	[187]
04		Pd	0.4 nm/%	5%	~ 30 seconds	[188]
05		Pd/WO_3	1.09 nm/%	0–1%		[189]
06		Pd		0.0–3.6%		[167]
07		WO_3/SIO_2	0.47 dB/%	0.000–0.901%		[172]
08		WO_3	2.074444 dB/%	0.00–1.32%		[190]
09		Pd	1.44 nm/%	8%	185 seconds	[164]
10	FPI	Pd	–0.09643 nm/% and –0.05 nm/%	0–8%	~ 2 minutes	[191]
11		Pt/WO_3	1 pm/ppm	0–20,916 ppm		[192]
12		Pt/WO_3	–5.1 nm/%	0.0–0.5%	< 1 minute	[193]
13		Pt/WO_3	> 1 pm/ppm	< 4%	< 1 minute	[173]
14		Pd-Y		0.0–5.5%		[194]
15		Pt-loaded WO_3/SiO_2	17.48 nm/%	0–4%		[195]
16		Pd/WO_3	1.26857 nm/%	0–10%		[196]
17		Pt-loaded WO_3/SiO_2	–1.04 nm/%	0.0–2.4%	~ 80 seconds	[197]
18		Pd/multilayer graphene		~20 ppm	~ 18 seconds	[178]
19		Pt-loaded WO_3	4.969 nm/%	0–4.%	120 seconds	[198]
20	SI	Pd/Ag composite	~ 131 pm/%	0–4%		[181]
21		Pt-loaded WO_3/SiO_2	–7.877 nm/%	0–1%	1 minute	[199]
22		Pt-loaded WO_3/SiO_2	–14.61 nm/%	0.0–0.8%		[184]

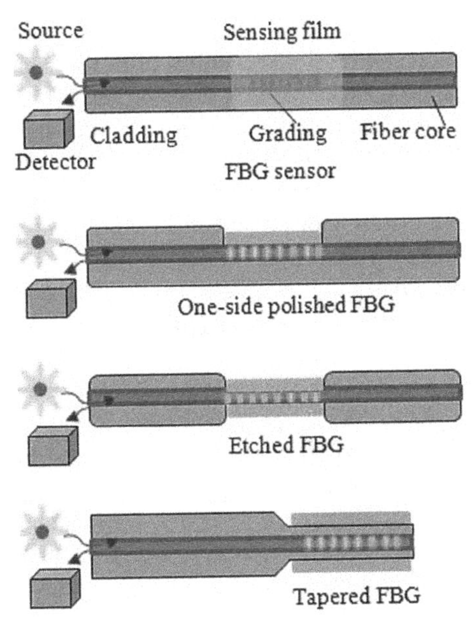

FIGURE 8.7 Schematic presentation of different types of FBG optical H_2 sensors.

oil [207]. Sensitivity of a sensor can be improved by using composite films instead of simple Pd [208]; hence, Caucheteur et al. (2008) independently used Pt-doped WO_3 films to develop FBG and FBG-superimposed LPG sensors, respectively, for the detection of H_2 [208]. Four years later, Yang et al. (2012) reported an FBG sensor based on the same film for the detection of H_2 up to 6.5% [209]. Another group, in an attempt to enhance sensor stability, incorporated Na^+ and K^+ into the α-MoO_3 nanobelt-based FBG sensor for the detection of H_2 [202].

One of the major challenges with the FBG sensors was to miniaturize them so that they could fit into portable equipment or household devices; this can increase their popularity hugely. In an attempt to do this, spiral micro FBGs were used, as these are very small sensors having very high sensitivity and specificity. Contributing to this, a spiral micro FBG sensor was fabricated using Pd/Ag film for the detection of H_2 up to 4% within 160–180 seconds [126]. Just a year later, Zhou et al. (2017) used a Pt-WO_3 composite film to devise a double-spiral micro FBG capable of detecting H_2 to 0.02–4.00% at 20 °C and 27% humidity; i.e. two to four times more sensitive than their earlier reported single-spiral FBG sensor [226].

Dai et al. (2013) devised the first etched FBG H_2 sensor with good reversibility by coating with a Pd Pd/Ag composite film [227]. The sensor showed that the wavelength shifts of FBG-20.6 µm, 38 µm, and 125 µm were 40 pm, 23 pm, and 8 pm, respectively, at 4% H_2. In another experiment the next year, Dai et al. (2014) demonstrated an etched FBG sensor with a stability of > 6 months, using Pd/Ni composite film and protected with polypropylene substrate [228]. Again in 2017, the group tried to explore the gasochromic effect of WO_3 to sense H_2 at less than 20 ppm by coating the sensor with WO_3-Pd_2Pt-Pt composite film [229].

To further improve the sensitivity of the sensors, dual-grating-based etched FBG H_2 sensors were developed with a detectability of 0–1%. One of the gratings was uncoated, and the other was coated with a layer of Pd of 150 nm but covered with teflon to protect from high humidity [217]. Further, a pressure-resistant FBG sensor was developed by Fisser et al. (2018) using micro twin-hole etched optical fiber containing double-grated silica films, one coated with Pd/Ti and the other with Pd foil attached [230]. The results showed higher sensitivity and accuracy with the foil sensor for detection of H_2.

A titled FBG (TFBG) sensor was fabricated to check H_2 leakage through cracks in transportation pipes by detecting the change in the refractive index of the Pd film on interaction with H_2 [224]. Very recently, another optical fiber-based TFBG H_2 sensor employing a Pd-Au alloy nanolayer was reported. The sensor exhibited a stabilization time of < 20 seconds and < 30 seconds during association and during dissociation phases, respectively [225]. Kurohiji et al. (2018) attempted to design a multipoint-leakage FBG sensor using Pt/SiO_2 catalyst film for the detection of H_2 at RT. Overall, FGA-based H_2 sensors are temperature-sensitive [129], cumbersome, and require high energy laser beams like ultraviolet lasers [162, 187], phase masks, and optical setups for grating [231]. Therefore, a multidisciplinary approach may be more fruitful in the near future. Different reported FBG sensors are listed in Table 8.3.

8.8 FUTURE PERSPECTIVES

8.8.1 Modification of Existing Sensors

The performance of a sensor can be determined by its size, simplicity, economic value, sensitivity, specificity, LOD, and stability. Specificity of the optical sensors needs further improvement for the detection of H_2 from a complicated gaseous mixture without being affected by the presence of other gases [52]. This will also minimize false positive or negative results. Further, simplicity and economic value of these sensors counts most for the use of these sensors in common household devices for the public, especially in developing countries like India [55]. Portability of any sensor plays a vital role in making it feasible for use in small equipment and remote devices. The sensitivity and LOD of an H_2 sensor determines its potential to detect the minimum possible H_2 with highest accuracy, and this should be improved further in future, especially for use in nuclear plants and for isotopic separations. Similarly, the stability of a sensor is among the most desirable qualities in terms of containing its cost [53]. Moreover, the focus must be on developing multidisciplinary H_2 sensors so as to widen their operational temperature, pressure, and pH [54]. Researchers must design more sensor arrays and multiplexed sensors for multi-location detection of H_2 in a gas supply and storage system.

8.8.2 Multidisciplinary Approaches

Raman is one of the simplest methods for directly detecting H_2, but to improve its sensitivity, photoacoustic stimulation using a different laser beam can be tested [108]. In an alternative approach, multiple-mirror systems can replace the conventional

TABLE 8.3
Different studies on FBG sensors reported for the detection of H_2 gas

Sl. No.	Sensing film	Film thickness	Sensitivity	Concentration of H_2 detected	Response time	Detection temperature (°C)	Reference
01	Pd	560 nm	1.95×10^{-2} nm/%	0.3–1.8%		RT	[210]
02	Pd	5 nm	8.4 pm/%	0.3–3.0%	10 seconds	RT	[211]
03	Pd	560 nm	60.73 pm/%	4%		RT	[212]
04	Pt-loaded WO_3			4%	10 seconds	RT (26)	[213]
05	Pd/Ag composite	110 nm		1.5–4.0%	280–300 seconds	RT	[214]
06	Pd/Ni composite	130 nm		0.5–4.0%	5–6 minutes	25	[215]
07	Pd	560 nm	0.042 (RT) and 0.044 pm/ppm (80 °C)	1,791.46 ppm		RT 80	[216]
08	Pd	150 nm	> 20 pm/%	0–1%	2–10 minutes		[217]
09	Pd/Ag	560 nm	0.477 (pm/(µL/L)	100–400 µl/L	4 hours 1 hour (60 ºC)	RT 60	[218]
10	Pt/SiO_2, Pt/WO_3 Pt/Fe_2O_3, Pt/ZnO, Pt/SnO_2,Pt/Al_2O_3		20–480 pm/%	0.1–1.0%	< 20 seconds	RT	[219]
11	Pt/WO_3		0.023 pm/ppm	15,000–20,000 ppm	60 seconds	RT	[220]
12	Pd-Ag	520 nm	25.5–51.5 pm/%	0.2–4.0%	160–180 seconds		[126]
13	WO_3-Pd_2Pt-Pt composite	WO_3: 200 nm Pd_2Pt: 20 nm Pt: 10 nm		10–100 ppm	9–680 seconds	25	[221]
14	Pt-WO_3	2 µm	522 pm/%	0.02–4.00%	10–30 seconds	RT	[222]
15	Pd	100 µm		5% and 1%	20–30 minutes	~ 90	[223]
16	Pd			1–4%	5 minutes		[224]
17	Pd	50 nm		0.5–2.0%	Association: < 20 seconds Dissociation:< 30 seconds		[225]

two-mirror Raman setup [112, 232]. The large size of the prism-based sensors make them difficult to use in remote sensing or other mini apparatus. Therefore, several localized SPR sensors were reported [124] using Pd coating, but the sensitivity of detection can further be increased by replacing the Pd film with Pd/WO_3, Pt/WO_3, Ni/Y, etc. composite films. These composites will also enhance the repeatability of the sensors, lowering the cost of detection many times over. Further, optical fiber sensors are small, stable, and sensitive, so are used often; but as illustrated by Wang et al. (2014), a multidisciplinary option like fiber grating interferometry optical fiber sensors should be explored in order to harness the synergistic sensing output in terms of sensitivity, LOD, and specificity [192]. According to several reports [52, 233], dual-grated FBG sensors have shown promising results, and to extend these studies, multiple-grated tapered or one-side polished fibers can be implemented in future. TFBG sensors have proven to be highly stable [225] and capable of detecting with high sensitivity H_2 leakage through the cracks in transportation pipes [224]. Collating this work with that of Kurohiji et al. (2018) [190], multipoint-leakage FBG sensors can be developed in future.

8.9 SUMMARY

In this chapter, we have discussed briefly the versatile of applications of H_2 and the risks associated with their leakage into the external environment. This highlights the need for a simple, sensitive, specific device for the quick detection of H_2 from any source. We focused on GC-based detection of H_2 and different detectors incorporating TC and MS for precise and sensitive detection of the gas. Different types of sensors have been reported by different research groups aiming to miniaturize the detecting devices and achieve a simple and quick detection method. The prime focus of the chapter was recent developments in optical methods for the detection of H_2. We have described methods like FTIR and Raman used for the direct detection of H_2 as well as indirect methods like SPR and optical fiber-based detection. The laser-based Raman stimulation to enhance the sensitivity of detection was also described. Under optical fiber-based detection, the sensors were categorized into different groups depending on their detection methods for the re-emitted light. The sensors were divided into direct intensity, SPR, interferometry, and FGA types.

The chapter concluded that although several types of H_2 sensors have been reported to date, optical sensors are the best because of their simplicity, specificity, and reproducibility. Still, many other modifications and advancements in design of these sensors are to be reported in the near future. This will further improve the sensitivity and specificity of detection, making it applicable in a wide range of environmental conditions without being affected by other gases. These sensors can be further developed for commercialization and practical utilization in various industries.

ACKNOWLEDGMENT

The authors would like to thank Dr. P. K. Dixit, Head of Department, Department of Zoology, Berhampur University, for providing timely insight and suggestions to complete the chapter.

REFERENCES

1. Esfandiar, A., Irajizad, A., Akhavan, O., Ghasemi, S., Gholami, M. R. 2014. Pd–WO_3/reduced graphene oxide hierarchical nanostructures as efficient hydrogen gas sensors. *International Journal of Hydrogen Energy*, *39* (15), 8169–8179. https://doi.org/10.1016/j.ijhydene.2014.03.117.
2. Dutta, S. 2014. A review on production, storage of hydrogen and its utilization as an energy resource. *Journal of Industrial and Engineering Chemistry*, *20* (4), 1148–1156. https://doi.org/10.1016/j.jiec.2013.07.037.
3. Salvi, B. L., Subramanian, K. A. 2015. Sustainable development of road transportation sector using hydrogen energy system. *Renewable and Sustainable Energy Reviews*, *51*, 1132–1155. https://doi.org/10.1016/j.rser.2015.07.030.
4. Gim, B.-J., Kim, J.-W., Choi, S.-J. 2005. The status of domestic hydrogen production, consumption, and distribution. *Transactions of the Korean Hydrogen and New Energy Society*, *16* (4), 391–399.
5. O'Hara, R., Holborn, P. G., Ingram, J. M., Ball, J., Rathbone, P., Edge, R. 2018. Chemochromic Pd-V_2O_5 Sensors for Passive Hydrogen Detection in Nuclear Containments. WM Symposia, Nuclear and Industrial Robotics, Remote Systems and Other Emerging Technologies. Phoenix, Arizona, 18–22 March.
6. Xiao, J., Travis, J. R., Breitung, W., Jordan, T. 2010. Numerical analysis of hydrogen risk mitigation measures for support of ITER licensing. *Fusion Engineering and Design*, *85* (2), 205–214. https://doi.org/10.1016/j.fusengdes.2009.12.008.
7. Abdelwahid, S., Nemitallah, M., Imteyaz, B., Abdelhafez, A., Habib, M. 2018. Effects of H_2 enrichment and inlet velocity on stability limits and shape of $CH_4/H_2–O_2/CO_2$ flames in a premixed swirl combustor. *Energy Fuels*, *32* (9), 9916–9925. https://doi.org/10.1021/acs.energyfuels.8b01958.
8. Wang, Y., Liu, X., Gu, M., An, X. 2019. Numerical simulation of the effects of hydrogen addition to fuel on the structure and soot formation of a laminar axisymmetric coflow $C_2H_4/(O_2\text{-}CO_2)$ diffusion flame. *Null*, *191* (10), 1743–1768. https://doi.org/10.1080/00102202.2018.1532413.
9. Xie, Y., Lv, N., Li, Q., Wang, J. 2020. Effects of CO addition on laminar flame characteristics and chemical reactions of H_2 and CH_4 in oxy-fuel (O_2/CO_2) atmosphere. *International Journal of Hydrogen Energy*, *45* (39), 20472–20481. https://doi.org/10.1016/j.ijhydene.2019.10.138.
10. Zappa, D. 2019. The influence of Nb on the synthesis of WO_3 nanowires and the effects on hydrogen sensing performance. *Sensors*, *19* (10), 2332. https://doi.org/10.3390/s19102332.
11. Kobayashi, H., Daimon, Y., Umemura, Y., Muto, D., Naruo, Y., Miyanabe, K. 2018. Temperature measurement and flow visualization of cryo-compressed hydrogen released into the atmosphere. *International Journal of Hydrogen Energy*, *43* (37), 17938–17953. https://doi.org/10.1016/j.ijhydene.2018.07.144.
12. Crowl, D. A., Jo, Y.-D. 2007. The hazards and risks of hydrogen. *Journal of Loss Prevention in the Process Industries*, *20* (2), 158–164. https://doi.org/10.1016/j.jlp.2007.02.002.
13. Nugroho, F. A. A., Darmadi, I., Cusinato, L., Susarrey-Arce, A., Schreuders, H., Bannenberg, L. J., da Silva Fanta, A. B., Kadkhodazadeh, S., Wagner, J. B., Antosiewicz, T. J., et al. 2019. Metal–polymer hybrid nanomaterials for plasmonic ultrafast hydrogen detection. *Nature Materials*, *18* (5), 489–495. https://doi.org/10.1038/s41563-019-0325-4.
14. Liu, Y., Chen, Y., Song, H., Zhang, G. 2013. Characteristics of an optical fiber hydrogen gas sensor based on a palladium and yttrium alloy thin film. *IEEE Sensors Journal*, *13* (7), 2699–2704. https://doi.org/10.1109/JSEN.2013.2258904.

15. Mohammadfam, I., Zarei, E. 2015. Safety risk modeling and major accidents analysis of hydrogen and natural gas releases: A comprehensive risk analysis framework. *International Journal of Hydrogen Energy, 40* (39), 13653–13663. https://doi.org/10.1016/j.ijhydene.2015.07.117.
16. Wang, B., Zhu, L. F., Yang, Y. H., Xu, N. S., Yang, G. W. 2008. Fabrication of a SnO_2 nanowire gas sensor and sensor performance for hydrogen. *The Journal of. Physical Chemistry C, 112* (17), 6643–6647. https://doi.org/10.1021/jp8003147.
17. Huang, X., Manolidis, M., Jun, S. C., Hone, J. 2005. Nanomechanical hydrogen sensing. *Applied Physics Letters, 86*, 143104/1–143104/3. https://doi.org/10.1063/1.1897445.
18. Jakubik, W. P. 2007. Investigations of thin film structures of WO_3 and WO_3 with Pd for hydrogen detection in a surface acoustic wave sensor system. *Thin Solid Films, 515* (23), 8345–8350. https://doi.org/10.1016/j.tsf.2007.03.024.
19. Baselt, D. R., Fruhberger, B., Klaassen, E., Cemalovic, S., Britton, C. L., Patel, S. V., Mlsna, T. E., McCorkle, D., Warmack, B. 2003. Design and performance of a microcantilever-based hydrogen sensor. *Sensors and Actuators B: Chemical, 88* (2), 120–131. https://doi.org/10.1016/S0925-4005(02)00315-5.
20. Liu, Y., Li, Y. 2019. Signal analysis and processing method of transmission optical fiber hydrogen sensors with multi-layer Pd–Y alloy films. *International Journal of Hydrogen Energy, 44* (49), 27151–27158. https://doi.org/10.1016/j.ijhydene.2019.08.143.
21. Bremer, K., Meinhardt-Wollweber, M., Thiel, T., Werner, G., Sun, T., Grattan, K. T. V., Roth, B. 2014. Sewerage tunnel leakage detection using a fibre optic moisture-detecting sensor system. *Sensors and Actuators A: Physical, 220*, 62–68. https://doi.org/10.1016/j.sna.2014.09.018.
22. Sakamoto, Y., Uemura, K., Ikuta, T., Maehashi, K. 2018. Palladium configuration dependence of hydrogen detection sensitivity based on graphene FET for breath analysis. *Japanese Journal of Applied Physics, 57* (4S), 04FP05. https://doi.org/10.7567/JJAP.57.04FP05.
23. Snavely, K., Subramaniam, B. 1998. Thermal conductivity detector analysis of hydrogen using helium carrier gas and HayeSep® D columns. *Journal of Chromatographic Science, 36* (4), 191–196. https://doi.org/10.1093/chromsci/36.4.191.
24. Chen, P., Fu, X., Hu, P., Xiao, C., Ren, X., Xia, X., Wang, H. 2017. Analysis of hydrogen isotopes by gas chromatography using a $MnCl_2$ coated γ-Al_2o_3 capillary packed column. *Se Pu, 35* (7), 766–771. https://doi.org/10.3724/sp.j.1123.2017.02028.
25. Falony, G., Verschaeren, A., Bruycker, F. D., Preter, V. D., Verbeke, K., Leroy, F., Vuyst, L. D. 2009. In vitro kinetics of prebiotic inulin-type fructan fermentation by butyrate-producing colon bacteria: implementation of online gas chromatography for quantitative analysis of carbon dioxide and hydrogen gas production. *Applied and Environmental Microbioliology, 75* (18), 5884–5892. https://doi.org/10.1128/AEM.00876-09.
26. Traore, S. I., Khelaifia, S., Armstrong, N., Lagier, J. C., Raoult, D. 2019. Isolation and culture of methanobrevibacter smithii by co-culture with hydrogen-producing bacteria on agar plates. *Clinical Microbiology and Infection, 25* (12), 1561.e1–1561.e5. https://doi.org/10.1016/j.cmi.2019.04.008.
27. He, Y.-S., Chen, K.-F., Lin, C.-H., Lin, M.-T., Chen, C.-C., Lin, C.-H. 2013. Use of an accelerometer and a microphone as gas detectors in the online quantitative detection of hydrogen released from ammonia borane by gas chromatography. *Analytical Chemistry, 85* (6), 3303–3308. https://doi.org/10.1021/ac303694j.
28. Cao, Z., Duan, X., Yao, P., Cui, W., Cheng, D., Zhang, J., Jin, Q., Chen, J., Dai, T., Shen, W. 2017. Hydrogen gas is involved in auxin-induced lateral root formation by modulating nitric oxide synthesis. *International Journal of Molecular Sciences, 18* (10), 2084. https://doi.org/10.3390/ijms18102084.

29. Fan, J., Wang, F., Sun, Q., Bin, F., Ding, J., Ye, H. 2017. SOFC detector for portable gas chromatography: High-sensitivity detection of dissolved gases in transformer oil. *IEEE Transactions on Dielectrics and Electrical Insulation, 24* (5), 2854–2863. https://doi.org/10.1109/TDEI.2017.006438.
30. Ganzha, V., Ivshin, K., Kammel, P., Kravchenko, P., Kravtsov, P., Petitjean, C., Trofimov, V., Vasilyev, A., Vorobyov, A., Vznuzdaev, M., et al. 2018. Measurement of trace impurities in ultra pure hydrogen and deuterium at the parts-per-billion level using gas chromatography. *Nuclear Instruments and Methods in Physics Research Section A: Accelerators, Spectrometers, Detectors and Associated Equipment, 880*, 181–187. https://doi.org/10.1016/j.nima.2017.10.096.
31. Abrahim, A., Cannavan, A., Kelly, S. D. 2020. Stable isotope analysis of non-exchangeable hydrogen in carbohydrates derivatised with *N*-methyl-bis-trifluoroacetamide by gas chromatography – chromium silver reduction/high temperature conversion-isotope ratio mass spectrometry (GC-CrAg/HTC-IRMS). *Food Chemistry, 318*, 126413. https://doi.org/10.1016/j.foodchem.2020.126413.
32. Maget, H. 1971. Electrochemical detection of H_2, CO, and hydrocarbons in inert or oxygen atmospheres. *Journal of The Electrochemical Society, 118*, 506–510. https://doi.org/10.1149/1.24080930.
33. Aroutiounian, V. 2007. Metal oxide hydrogen, oxygen, and carbon monoxide sensors for hydrogen setups and cells. *International Journal of Hydrogen Energy, 32* (9), 1145–1158. https://doi.org/10.1016/j.ijhydene.2007.01.004.
34. Yang, F., Zhao, Y., Qi, Y., Tan, Y. Z., Ho, H. L., Jin, W. 2019. Towards label-free distributed fiber hydrogen sensor with stimulated Raman spectroscopy. *Optics Express, 27* (9), 12869–12882. https://doi.org/10.1364/OE.27.012869.
35. Toonen, A., van Loon, R. 2013. *Hydrogen Detection with a TCD Using Mixed Carrier Gas on the Agilent Micro GC.* Agilent Technologies. www.agilent.com/cs/library/applications/5991-3199EN.pdf.
36. Harkinezhad, B., Soleimani, A., Hossein-Babaei, F. 2019. Hydrogen level detection via thermal conductivity measurement using temporal temperature monitoring. In *2019 27th Iranian Conference on Electrical Engineering (ICEE)*, Yazd, Iran: IEEE, 408–411. https://doi.org/10.1109/IranianCEE.2019.8786730.
37. Berndt, D., Muggli, J., Wittwer, F., Langer, C., Heinrich, S., Knittel, T., Schreiner, R. 2020. MEMS-based thermal conductivity sensor for hydrogen gas detection in automotive applications. *Sensors and Actuators A: Physical, 305*, 111670. https://doi.org/10.1016/j.sna.2019.111670.
38. Chen, S. 2010. *Surface-Micromachined Thermal Conductivity Gas Sensors For Hydrogen Detection*, Delft University of Technology.
39. Hübert, T., Boon-Brett, L., Palmisano, V., Bader, M. A. 2014. Developments in gas sensor technology for hydrogen safety. *International Journal of Hydrogen Energy, 39* (35), 20474–20483. https://doi.org/10.1016/j.ijhydene.2014.05.042.
40. Ellefson, R. E., Moddeman, W. E., Dylla, H. F. 1981. Hydrogen isotope analysis by quadrupole mass spectrometry. *Journal of Vacuum Science and Technology, 18* (3), 1062–1067. https://doi.org/10.1116/1.570885.
41. Genty, C., Schott, R. 1970. Quantitative analysis for the isotopes of hydrogen-H_2,HD,HT,D_2,DT, and T_2-by gas chromatography. *Analytical Chemistry, 42* (1), 7–11. https://doi.org/10.1021/ac60283a039.
42. Uda, T., Okuno, K., Suzuki, T., Naruse, Y. 1991. Gas chromatography for measurement of hydrogen isotopes at tritium processing. *Journal of Chromatography A, 586* (1), 131–137. https://doi.org/10.1016/0021-9673(91)80030-K.

43. Gehre, M., Renpenning, J., Gilevska, T., Qi, H., Coplen, T. B., Meijer, H. A. J., Brand, W. A., Schimmelmann, A. 2015. On-line hydrogen-isotope measurements of organic samples using elemental chromium: An extension for high temperature elemental-analyzer techniques. *Analytical Chemistry*, *87* (10), 5198–5205. https://doi.org/10.1021/acs.analchem.5b00085.
44. Nair, S., Geilmann, H., Coplen, T. B., Qi, H., Gehre, M., Schimmelmann, A., Brand, W. A. 2015. Isotopic disproportionation during hydrogen isotopic analysis of nitrogen-bearing organic compounds. *Rapid Communications in Mass Spectrometry*, *29* (9), 878–884. https://doi.org/10.1002/rcm.7174.
45. Gehre, M., Renpenning, J., Geilmann, H., Qi, H., Coplen, T. B., Kümmel, S., Ivdra, N., Brand, W. A., Schimmelmann, A. 2017. Optimization of on-line hydrogen stable isotope ratio measurements of halogen- and sulfur-bearing organic compounds using elemental analyzer-chromium/high-temperature conversion isotope ratio mass spectrometry (EA-Cr/HTC-IRMS). *Rapid Communications in Mass Spectrometry*, *31* (6), 475–484. https://doi.org/10.1002/rcm.7810.
46. Eysseric, E., Bellerose, X., Lavoie, J.-M., Segura, P. A. 2016. Post-column hydrogen–deuterium exchange technique to assist in the identification of small organic molecules by mass spectrometry. *Canadian Journal of Chemistry*, *94* (9). https://doi.org/10.1139/cjc-2016-0281.
47. Takai, K., Chiba, Y., Noguchi, K., Nozue, A. 2002. Visualization of the hydrogen desorption process from ferrite, pearlite, and graphite by secondary ion mass spectrometry. *Metallurgical and Materials Transactions A*, *33* (8), 2659–2665. https://doi.org/10.1007/s11661-002-0387-8.
48. Awane, T., Fukushima, Y., Matsuo, T., Matsuoka, S., Murakami, Y., Miwa, S. 2011. Highly sensitive detection of net hydrogen charged into austenitic stainless steel with secondary ion mass spectrometry. *Analytical Chemistry*, *83* (7), 2667–2676. https://doi.org/10.1021/ac103100b.
49. Kamiński, M., Kartanowicz, R., Jastrzębski, D., Kamiński, M. M. 2003. Determination of carbon monoxide, methane and carbon dioxide in refinery hydrogen gases and air by gas chromatography. *Journal of Chromatography A*, *989* (2), 277–283. https://doi.org/10.1016/S0021-9673(03)00032-3.
50. Camargo, F. P., Sarti, A., Alécio, A. C., Sabatini, C. A., Adorno, M. A. T., Duarte, I. C. S., Varesche, M. B. A., Camargo, F. P., Sarti, A., Alécio, A. C. et al. 2020. Limonene quantification by gas chromatography with mass spectrometry (GC-MS) and its effects on hydrogen and volatile fatty acids production in anaerobic reactors. *Química Nova*, *43* (7), 844–850. https://doi.org/10.21577/0100-4042.20170557.
51. Tang, C., Tan, J., Fan, Y., Peng, X. 2020. Ascertaining hydrogen-abstraction reaction efficiencies of halogenated organic compounds in electron impact ionization processes by gas chromatography–high-resolution mass spectrometry. *ACS Omega*, *5* (15), 8496–8507. https://doi.org/10.1021/acsomega.9b03895.
52. Fisser, M., Badcock, R. A., Teal, P. D., Hunze, A. 2018. Improving the sensitivity of palladium-based fiber optic hydrogen sensors. *Journal of Lightwave Technology*, *36* (11), 2166–2174. https://doi.org/10.1109/JLT.2018.2807789.
53. Zhong, X., Yang, M., Huang, C., Wang, G., Dai, J., Bai, W. 2016. Water photolysis effect on the long-term stability of a fiber optic hydrogen sensor with Pt/WO_3. *Scientific Reports*, *6* (1), 39160. https://doi.org/10.1038/srep39160.
54. Wang, B., Zhu, Y., Chen, Y., Song, H., Huang, P., Dao, D. V. 2017. Hydrogen sensor based on palladium-yttrium alloy nanosheet. *Materials Chemistry and Physics*, *194*, 231–235. https://doi.org/10.1016/j.matchemphys.2017.03.042.

55. Hübert, T. 2016. *H2Sense – Cost-Effective and Reliable Hydrogen Sensors for Facilitating the Safe Use of Hydrogen*. European Commission.
56. Hübert, T., Boon-Brett, L., Black, G., Banach, U. 2011. Hydrogen sensors: A review. *Sensors and Actuators B: Chemical, 157* (2), 329–352. https://doi.org/10.1016/j.snb.2011.04.070.
57. Cheng, Y.-T., Li, Y., Lisi, D., Wang, W. M. 1996. Preparation and characterization of Pd/Ni thin films for hydrogen sensing. *Sensors and Actuators B: Chemical, 30* (1), 11–16. https://doi.org/10.1016/0925-4005(95)01734-D.
58. Wang, M., Feng, Y. 2007. Palladium–silver thin film for hydrogen sensing. *Sensors and Actuators B: Chemical, 123* (1), 101–106. https://doi.org/10.1016/j.snb.2006.07.030.
59. Huang, X. M. H., Manolidis, M., Jun, S. C., Hone, J. 2005. Nanomechanical hydrogen sensing. *Applied Physics Letters, 86* (14), 143104. https://doi.org/10.1063/1.1897445.
60. De, G., Köhn, R., Xomeritakis, G., Brinker, C. J. 2007. Nanocrystalline mesoporous palladium activated tin oxide thin films as room-temperature hydrogen gas sensors. *Chemical Communications*, 18, 1840–1842. https://doi.org/10.1039/B700029D.
61. Shukla, S., Patil, S., Kuiry, S. C., Rahman, Z., Du, T., Ludwig, L., Parish, C., Seal, S. 2003. Synthesis and characterization of sol–gel derived nanocrystalline tin oxide thin film as hydrogen sensor. *Sensors and Actuators B: Chemical, 96* (1), 343–353. https://doi.org/10.1016/S0925-4005(03)00568-9.
62. Rout, C. S., Raju, A. R., Govindaraj, A., Rao, C. N. R. 2006. Hydrogen sensors based on ZnO nanoparticles. *Solid State Communications, 138* (3), 136–138. https://doi.org/10.1016/j.ssc.2006.02.016.
63. Varghese, O. K., Gong, D., Paulose, M., Ong, K. G., Grimes, C. A. 2003. Hydrogen sensing using titania nanotubes. *Sensors and Actuators B: Chemical, 93* (1), 338–344. https://doi.org/10.1016/S0925-4005(03)00222-3.
64. Chaudhari, G. N., Bende, A. M., Bodade, A. B., Patil, S. S., Sapkal, V. S. 2006. Structural and gas sensing properties of nanocrystalline TiO_2:WO_3-based hydrogen sensors. *Sensors and Actuators B: Chemical, 115* (1), 297–302. https://doi.org/10.1016/j.snb.2005.09.014.
65. Matushko, I. P., Yatsimirskii, V. K., Maksimovich, N. P., Nikitina, N. V., Silenko, P. M., Ruchko, V. P., Ishchenko, V. B. 2008. Sensitivity to hydrogen of sensor materials based on SnO_2 promoted with 3D metals. *Theoretical and Experimental Chemistry, 44* (2), 128–133. https://doi.org/10.1007/s11237-008-9008-y.
66. Panchapakesan, B., Cavicchi, R., Semancik, S., DeVoe, D. L. 2005. Sensitivity, selectivity and stability of tin oxide nanostructures on large area arrays of microhotplates. *Nanotechnology, 17* (2), 415–425. https://doi.org/10.1088/0957-4484/17/2/012.
67. Sun, Y., Wang, H. H. 2007. High-performance, flexible hydrogen sensors that use carbon nanotubes decorated with palladium nanoparticles. *Advanced Materials, 19* (19), 2818–2823. https://doi.org/10.1002/adma.200602975.
68. Krishna Kumar, M., Ramaprabhu, S. 2007. Palladium dispersed multiwalled carbon nanotube based hydrogen sensor for fuel cell applications. *International Journal of Hydrogen Energy, 32* (13), 2518–2526. https://doi.org/10.1016/j.ijhydene.2006.11.015.
69. Tsai, T.-H., Chen, H.-I., Lin, K.-W., Hung, C.-W., Hsu, C.-H., Chen, L.-Y., Chu, K.-Y., Liu, W.-C. 2008. Comprehensive study on hydrogen sensing properties of a Pd–AlGaN-based Schottky diode. *International Journal of Hydrogen Energy, 33* (12), 2986–2992. https://doi.org/10.1016/j.ijhydene.2008.03.055.
70. Miyoshi, M., Kuraoka, Y., Asai, K., Shibata, T., Tanaka, M., Egawa, T. 2007. Electrical characterization of Pt/AlGaN/GaN Schottky diodes grown using AlN template and

their application to hydrogen gas sensors. *Journal of Vacuum Science & Technology B: Microelectronics and Nanometer Structures Processing, Measurement, and Phenomena, 25* (4), 1231–1235. https://doi.org/10.1116/1.2749530.

71. Pandis, C., Brilis, N., Bourithis, E., Tsamakis, D., Ali, H., Krishnamoorthy, S., Iliadis, A. A., Kompitsas, M. 2007. Low–temperature hydrogen sensors based on Au nanoclusters and Schottky contacts on ZnO films deposited by pulsed laser deposition on Si and SiO_2 substrates. *IEEE Sensors Journal, 7* (3), 448–454. https://doi.org/10.1109/JSEN.2007.891944.
72. Weidemann, O., Hermann, M., Steinhoff, G., Wingbrant, H., Lloyd Spetz, A., Stutzmann, M., Eickhoff, M. 2003. Influence of surface oxides on hydrogen-sensitive Pd:GaN Schottky diodes. *Applied Physics Letters, 83* (4), 773–775. https://doi.org/10.1063/1.1593794.
73. Nakagomi, S., Shida, T., Hoshi, H., Kokubun, Y. 2007. Field-effect hydrogen sensor device with floating gate exhibiting unique behavior. *Sensors and Actuators B: Chemical, 125* (2), 408–414. https://doi.org/10.1016/j.snb.2007.02.034.
74. Kandasamy, S., Wlodarski, W., Holland, A., Nakagomi, S., Kokubun, Y. 2007. Electrical characterization and hydrogen gas sensing properties of a N-ZnO/p-SiC Pt-gate metal semiconductor field effect transistor. *Applied Physics Letters, 90* (6), 064103. https://doi.org/10.1063/1.2450668.
75. Cheng, C.-C., Tsai, Y.-Y., Lin, K.-W., Chen, H.-I., Hsu, W.-H., Hong, C.-W., Liu, W.-C. 2006. Characteristics of a Pd–oxide–$In_{0.49}Ga_{0.51}P$ high electron mobility transistor (HEMT)-based hydrogen sensor. *Sensors & Actuators: B. Chemical, 1* (113), 29–35. https://doi.org/10.1016/j.snb.2005.02.019.
76. Hung, C., Chang, H.-C., Tsai, Y.-Y., Lai, P.-H., Fu, S., Chen, T., Chen, H., Liu, W. 2007. Study of a new field-effect resistive hydrogen sensor based on a Pd/oxide/AlGaAs transistor. *IEEE Transactions on Electron Devices, 54* (5), 1224–1231. https://doi.org/10.1109/TED.2007.893813.
77. Yamaguchi, T., Kiwa, T., Tsukada, K., Yokosawa, K. 2007. Oxygen interference mechanism of platinum–FET hydrogen gas sensor. *Sensors and Actuators A: Physical, 136* (1), 244–248. https://doi.org/10.1016/j.sna.2006.11.026.
78. Lu, C., Chen, Z., Saito, K. 2007. Hydrogen sensors based on Ni/SiO_2/Si MOS capacitors. *Sensors and Actuators B: Chemical, 122* (2), 556–559. https://doi.org/10.1016/j.snb.2006.06.029.
79. Yadav, L., Chandra Gupta, N., Dwivedi, R., Singh, R. S. 2007. Sensing behavior and mechanism of titanium dioxide-based MOS hydrogen sensor. *Microelectronics Journal, 38* (12), 1226–1232. https://doi.org/10.1016/j.mejo.2007.09.020.
80. Astbury, G. R., Hawksworth, S. J. 2007. Spontaneous ignition of hydrogen leaks: A review of postulated mechanisms. *International Journal of Hydrogen Energy, 32* (13), 2178–2185. https://doi.org/10.1016/j.ijhydene.2007.04.005.
81. Sadek, A. Z., Wlodarski, W., Shin, K., Kaner, R. B., Kalantar-zadeh, K. A. 2008. Polyaniline/WO_3 nanofiber composite-based ZnO/64° YX $LiNbO_3$ SAW hydrogen gas sensor. *Synthetic Metals, 158* (1), 29–32. https://doi.org/10.1016/j.synthmet.2007.11.008.
82. Sadek, A. Z., Wlodarski, W., Li, Y. X., Yu, W., Li, X., Yu, X., Kalantar-zadeh, K. 2007. A ZnO nanorod based layered ZnO/64° YX $LiNbO_3$ SAW hydrogen gas sensor. *Thin Solid Films, 515* (24), 8705–8708. https://doi.org/10.1016/j.tsf.2007.04.009.
83. Ippolito, S. J., Kandasamy, S., Kalantar-zadeh, K., Wlodarski, W., Holland, A. 2005. Comparison between conductometric and layered surface acoustic wave hydrogen gas sensors. *Smart Materials and Structures, 15* (1), S131–S136. https://doi.org/10.1088/0964-1726/15/1/021.

84. Jakubik, W. P. 2007. Investigations of thin film structures of WO_3 and WO_3 with Pd for hydrogen detection in a surface acoustic wave sensor system. *Thin Solid Films, 515* (23), 8345–8350. https://doi.org/10.1016/j.tsf.2007.03.024.
85. Tajima, K., Choi, Y., Shin, W., Izu, N., Matsubara, I., Murayama, N. 2006. Micro-thermoelectric hydrogen sensors with Pt thin film and Pt/alumina thick film catalysts. *Journal of The Electrochemical Society, 153* (3), H58. https://doi.org/10.1149/1.2165777.
86. Arndt, M. 2002. Micromachined thermal conductivity hydrogen detector for automotive applications. In *Proceedings of IEEE Sensors*, Vol. 2, pp 1571–1575. IEEE: Orlando, FL. https://doi.org/10.1109/ICSENS.2002.1037357.
87. Leonardi, S. G., Bonavita, A., Donato, N., Neri, G. 2018. Development of a hydrogen dual sensor for fuel cell applications. *International Journal of Hydrogen Energy, 43* (26), 11896–11902. https://doi.org/10.1016/j.ijhydene.2018.02.019.
88. Romanelli, G., Rudić, S., Zanetti, M., Andreani, C., Fernandez-Alonso, F., Gorini, G., Krzystyniak, M., Škoro, G. 2018. Measurement of the para-hydrogen concentration in the ISIS moderators using neutron transmission and thermal conductivity. *Nuclear Instruments and Methods in Physics Research Section A: Accelerators, Spectrometers, Detectors and Associated Equipment, 888*, 88–95. https://doi.org/10.1016/j.nima.2018.01.039.
89. Tardy, P., Coulon, J.-R., Lucat, C., Menil, F. 2004. Dynamic thermal conductivity sensor for gas detection. *Sensors and Actuators B: Chemical, 98* (1), 63–68. https://doi.org/10.1016/j.snb.2003.09.019.
90. Li, F., Shang, F., Cheng, S., Ma, W., Jin, H., Zhang, X., Guo, L. 2020. Thermal conductivity measurements of the H_2/CO_2 mixture using the short-hot-wire method at 323.15–620.05 K and 2.14–9.37 Mpa. *International Journal of Hydrogen Energy, 45* (55), 31213–31224. https://doi.org/10.1016/j.ijhydene.2020.08.023.
91. Simon, I., Arndt, M. 2002. Thermal and gas-sensing properties of a micromachined thermal conductivity sensor for the detection of hydrogen in automotive applications. *Sensors and Actuators A: Physical*, 97–98, 104–108. https://doi.org/10.1016/S0924-4247(01)00825-1.
92. Pak, Y., Lim, N., Kumaresan, Y., Lee, R., Kim, K., Kim, T. H., Kim, S.-M., Kim, J. T., Lee, H., Ham, M.-H., et al. 2015. Palladium nanoribbon array for fast hydrogen gas sensing with ultrahigh sensitivity. *Advanced Materials, 27* (43), 6945–6952. https://doi.org/10.1002/adma.201502895.
93. Zhao, D., Brown, A., Wang, T., Yoshizawa, S., Sfeir, C., Heineman, W. R. 2018. In vivo quantification of hydrogen gas concentration in bone marrow surrounding magnesium fracture fixation hardware using an electrochemical hydrogen gas sensor. *Acta Biomater, 73*, 559–566. https://doi.org/10.1016/j.actbio.2018.04.032.
94. Arora, K., Srivastava, S., Solanki, P. R., Puri, N. K. 2019. Electrochemical hydrogen gas sensing employing palladium oxide/reduced graphene oxide (PdO-RGO) nanocomposites. *IEEE Sensors Journal, 19* (18), 8262–8271. https://doi.org/10.1109/JSEN.2019.2918360.
95. Holstein, N., Krauss, W., Konys, J., Nitti, F. S. 2019. Development of an electrochemical sensor for hydrogen detection in liquid lithium for IFMIF-DONES. *Fusion Engineering and Design, 146*, 1441–1445. https://doi.org/10.1016/j.fusengdes.2019.02.100.
96. Mashao, G., Ramohlola, K. E., Mdluli, S. B., Monama, G. R., Hato, M. J., Makgopa, K., Molapo, K. M., Ramoroka, M. E., Iwuoha, E. I., Modibane, K. D. 2019. Zinc-based zeolitic benzimidazolate framework/polyaniline nanocomposite for electrochemical sensing of hydrogen gas. *Materials Chemistry and Physics, 230*, 287–298. https://doi.org/10.1016/j.matchemphys.2019.03.079.

97. Khisina, N. R., Wirth, R., Andrut, M., Ukhanov, A. V. 2001. Extrinsic and intrinsic mode of hydrogen occurrence in natural olivines: FTIR and TEM investigation. *Physics and Chemistry of Minerals*, *28* (5), 291–301. https://doi.org/10.1007/s002690100162.
98. Koga, K., Hauri, E., Hirschmann, M., Bell, D. 2003. Hydrogen concentration analyses using SIMS and FTIR: Comparison and calibration for nominally anhydrous minerals. *Geochemistry, Geophysics, Geosystems*, *4* (2). https://doi.org/10.1029/2002GC000378.
99. Aubaud, C., Withers, A. C., Hirschmann, M. M., Guan, Y., Leshin, L. A., Mackwell, S. J., Bell, D. R. 2007. Intercalibration of FTIR and SIMS for hydrogen measurements in glasses and nominally anhydrous minerals. *American Mineralogist*, *92* (5–6), 811–828. https://doi.org/10.2138/am.2007.2248.
100. Pan, Y., Mei, D., Liu, C., Ge, Q. 2011. Hydrogen adsorption on Ga_2O_3 surface: A combined experimental and computational study. *The Journal of Physical Chemistry C*, *115* (20), 10140–10146. https://doi.org/10.1021/jp2014226.
101. Panayotov, D., Ivanova, E., Mihaylov, M., Chakarova, K., Spassov, T., Hadjiivanov, K. 2015. Hydrogen spillover on Rh/TiO_2: The FTIR study of donated electrons, co-adsorbed CO and H/D exchange. *Physical Chemistry Chemical Physics*, *17* (32), 20563–20573. https://doi.org/10.1039/C5CP03148F.
102. Enders, M., Kleber, M., Derscheid, G., Hofmann, K., Bauer, H.-D., Scheppat, B. 2020. Parallel FTIR-ATR spectroscopy and gravimetry for the in situ hydrogen desorption measurement of $NaAlH_4$ powder compacts. *Applied Optics*, *59* (30), 9510–9519. https://doi.org/10.1364/AO.403846.
103. Adamopoulos, G., Robertson, J., Morrison, N. A., Godet, C. 2004. Hydrogen content estimation of hydrogenated amorphous carbon by visible Raman spectroscopy. *Journal of Applied Physics*, *96* (11), 6348–6352. https://doi.org/10.1063/1.1811397.
104. Sturm, M., Schlösser, M., Lewis, R. J., Bornschein, B., Drexlin, G., Telle, H. H. 2010. Monitoring of all hydrogen isotopologues at tritium laboratory karlsruhe using Raman spectroscopy. *Laser Physics*, *20* (2), 493–507. https://doi.org/10.1134/S1054660X10030163.
105. Numata, Y., Shinohara, Y., Kitayama, T., Tanaka, H. 2013. Rapid and accurate quantitative analysis of fermentation gases by Raman spectroscopy. *Process Biochemistry*, *48* (4), 569–574. https://doi.org/10.1016/j.procbio.2013.02.018.
106. Gao, Y., Dai, L.-K., Zhu, H.-D., Chen, Y.-L., Zhou, L. 2019. Quantitative analysis of main components of natural gas based on Raman spectroscopy. *Chinese Journal of Analytical Chemistry*, *47* (1), 67–76. https://doi.org/10.1016/S1872-2040(18)61135-1.
107. Li, J., Chou, I.-M. 2015. Hydrogen in silicate melt inclusions in quartz from granite detected with Raman spectroscopy. *Journal of Raman Spectroscopy*, *46* (10), 983–986. https://doi.org/10.1002/jrs.4644.
108. Sharma, R., Poonacha, S., Bekal, A., Vartak, S., Weling, A., Tilak, V., Mitra, C. 2016. Raman analyzer for sensitive natural gas composition analysis. *Optical Engineering*, *55* (10), 104103. https://doi.org/10.1117/1.OE.55.10.104103.
109. Tedder, S. A., Wheeler, J. L., Cutler, A. D., Danehy, P. M. 2010. Width-increased dual-pump enhanced coherent anti-Stokes Raman spectroscopy. *Applied Optics*, *49* (8), 1305–1313. https://doi.org/10.1364/AO.49.001305.
110. Danehy, P. M., Bathel, B. F., Johansen, C. T., Winter, M., O'Byrne, S., Cutler, A. D. 2015. Molecular-based optical diagnostics for hypersonic nonequilibrium flows. In *Hypersonic Nonequilibrium Flows: Fundamentals and Recent Advances*, pp 343–470. American Institute of Aeronautics and Astronautics, Inc. https://doi.org/10.2514/5.9781624103292.0343.0470.
111. Spencer, C. L., Watson, V., Hippler, M. 2012. Trace gas detection of molecular hydrogen H_2 by photoacoustic stimulated Raman spectroscopy (PARS). *Analyst*, *137* (6), 1384–1388. https://doi.org/10.1039/C2AN15990B.

112. Wen, C., Huang, X., Wang, W., Shen, C., Li, H. 2019. Multiple-pass-enhanced Raman spectroscopy for long-term monitoring of hydrogen isotopologues. *Journal of Raman Spectroscopy*, *50* (10), 1555–1560. https://doi.org/10.1002/jrs.5649.
113. Wei, X., Wei, T., Xiao, H., Lin, Y. S. 2008. Nano-structured Pd-long period fiber gratings integrated optical sensor for hydrogen detection. *Sensors and Actuators B: Chemical*, *134* (2), 687–693. https://doi.org/10.1016/j.snb.2008.06.018.
114. Hanf, S., Bögözi, T., Keiner, R., Frosch, T., Popp, J. 2015. Fast and highly sensitive fiber-enhanced Raman spectroscopic monitoring of molecular H_2 and CH_4 for point-of-care diagnosis of malabsorption disorders in exhaled human breath. *Analytical Chemistry*, *87* (2), 982–988. https://doi.org/10.1021/ac503450y.
115. Knebl, A., Yan, D., Popp, J., Frosch, T. 2018. Fiber enhanced Raman gas spectroscopy. *TrAC Trends in Analytical Chemistry*, *103*, 230–238. https://doi.org/10.1016/j.trac.2017.12.001.
116. Qi, Y., Zhao, Y., Bao, H., Jin, W., Ho, H. L. 2019. Nanofiber enhanced stimulated Raman spectroscopy for ultra-fast, ultra-sensitive hydrogen detection with ultra-wide dynamic range. *Optica*, *6* (5), 570–576. https://doi.org/10.1364/OPTICA.6.000570.
117. Yang, L., Chen, L., Chen, Y.-C., Kang, L., Yu, J., Wang, Y., Lu, C., Mashimo, T., Yoshiasa, A., Lin, C.-H. 2019. Homogeneously alloyed nanoparticles of immiscible Ag-Cu with ultrahigh antibacterial activity. *Colloids Surf B Biointerfaces*, *180*, 466–472. https://doi.org/10.1016/j.colsurfb.2019.05.018.
118. Chadwick, B., Gal, M. 1993. Enhanced optical detection of hydrogen using the excitation of surface plasmons in palladium. *Applied Surface Science*, *68* (1), 135–138. https://doi.org/10.1016/0169-4332(93)90222-W.
119. Chadwick, B., Tann, J., Brungs, M., Gal, M. 1994. A hydrogen sensor based on the optical generation of surface plasmons in a palladium alloy. *Sensors and Actuators B: Chemical*, *17* (3), 215–220. https://doi.org/10.1016/0925-4005(93)00875-Y.
120. Liu, H., Wang, M., Wang, Q., Li, H., Ding, Y., Zhu, C. 2018. Simultaneous measurement of hydrogen and methane based on PCF-SPR structure with compound film-coated side-holes. *Optical Fiber Technology*, *45*, 1–7. https://doi.org/10.1016/j.yofte.2018.05.007.
121. Cavalcanti, G. O., Oliveira, S. C., Fontana, E., Azevedo, A. 2009. Wavelength and Pd thickness optimization for SPR-based hydrogen sensors. In *Frontiers in Optics 2009/ Laser Science XXV/Fall 2009 OSA Optics & Photonics Technical Digest*, Paper FWT3. Optical Society of America. https://doi.org/10.1364/FIO.2009.FWT3.
122. Bhatia, P., Gupta, B. D. 2012. Surface plasmon resonance based fiber optic hydrogen sensor utilizing wavelength interrogation. In *Third Asia Pacific Optical Sensors Conference*, International Society for Optics and Photonics, Vol. 8351, p 83511V. https://doi.org/10.1117/12.914046.
123. Watkins, W. L., Borensztein, Y. 2018. Ultrasensitive and fast single wavelength plasmonic hydrogen sensing with anisotropic nanostructured Pd films. *Sensors and Actuators B: Chemical*, *273*, 527–535. https://doi.org/10.1016/j.snb.2018.06.013.
124. Shafieyan, A. R., Ranjbar, M., Kameli, P. 2019. Localized surface plasmon resonance H_2 detection by MoO_3 colloidal nanoparticles fabricated by the flame synthesis method. *International Journal of Hydrogen Energy*, *44* (33), 18628–18638. https://doi.org/10.1016/j.ijhydene.2019.05.171.
125. Wang, X., Wolfbeis, O. S. 2020. Fiber-optic chemical sensors and biosensors (2015–2019). *Analytical Chemistry*, *92* (1), 397–430. https://doi.org/10.1021/acs.analchem.9b04708.
126. Zhou, X., Dai, Y., Zou, M., Karanja, J. M., Yang, M. 2016. FBG hydrogen sensor based on spiral microstructure ablated by femtosecond laser. *Sensors and Actuators B: Chemical*, *236*, 392–398. https://doi.org/10.1016/j.snb.2016.06.027.
127. Tan, Y. C., Ji, W. B., Mamidala, V., Chow, K. K., Tjin, S. C. 2014. Carbon-nanotube-deposited long period fiber grating for continuous refractive index sensor applications.

Sensors and Actuators B: Chemical, *196*, 260–264. https://doi.org/10.1016/j.snb.2014.01.063.

128. Arasu, P. T., Noor, A. S. M., Khalaf, A. L., Yaacob, M. H. 2016. Highly sensitive plastic optical fiber with palladium sensing layer for detection of hydrogen gas. In *2016 IEEE Region 10 Symposium (TENSYMP)*, pp 390–393. https://doi.org/10.1109/TENCONSpring.2016.7519438.
129. Hosoki, A., Nishiyama, M., Igawa, H., Seki, A., Choi, Y., Watanabe, K. 2013. A surface plasmon resonance hydrogen sensor using Au/Ta_2O_5/Pd multi-layers on hetero-core optical fiber structures. *Sensors and Actuators B: Chemical*, *185*, 53–58. https://doi.org/10.1016/j.snb.2013.04.072.
130. Javahiraly, N. 2015. Review on hydrogen leak detection: Comparison between fiber optic sensors based on different designs with palladium. *Optical Engineering*, *54* (3), 030901. https://doi.org/10.1117/1.OE.54.3.030901.
131. Zhang, Y., Peng, H., Qian, X., Zhang, Y., An, G., Zhao, Y. 2017. Recent advancements in optical fiber hydrogen sensors. *Sensors and Actuators B: Chemical*, *244*, 393–416. https://doi.org/10.1016/j.snb.2017.01.004.
132. Gullapalli, S. K., Vemuri, R. S., Ramana, C. V. 2010. Structural transformation induced changes in the optical properties of nanocrystalline tungsten oxide thin films. *Applied Physics Letters*, *96* (17), 171903. https://doi.org/10.1063/1.3421540.
133. Zheng, H., Ou, J. Z., Strano, M. S., Kaner, R. B., Mitchell, A., Kalantar-zadeh, K. 2011. Nanostructured tungsten oxide – properties, synthesis, and applications. *Advanced Functional Materials*, *21* (12), 2175–2196. https://doi.org/10.1002/adfm.201002477.
134. Shanak, H., Schmitt, H., Nowoczin, J., Ziebert, C. 2004. Effect of Pt-catalyst on gasochromic WO_3 films: Optical, electrical and AFM investigations. *Solid State Ionics*, *171* (1), 99–106. https://doi.org/10.1016/j.ssi.2004.04.001.
135. Boudiba, A., Roussel, P., Zhang, C., Olivier, M.-G., Snyders, R., Debliquy, M. 2013. Sensing mechanism of hydrogen sensors based on palladium-loaded tungsten oxide (Pd–WO_3). *Sensors and Actuators B: Chemical*, *187*, 84–93. https://doi.org/10.1016/j.snb.2012.09.063.
136. Correia, R., James, S., Lee, S.-W., Morgan, S. P., Korposh, S. 2018. Biomedical application of optical fibre sensors. *Journal of Optics*, *20* (7), 073003. https://doi.org/10.1088/2040-8986/aac68d.
137. Tabib-Azar, M., Sutapun, B., Petrick, R., Kazemi, A. 1999. Highly sensitive hydrogen sensors using palladium coated fiber optics with exposed cores and evanescent field interactions. *Sensors and Actuators B: Chemical*, *56* (1), 158–163. https://doi.org/10.1016/S0925-4005(99)00177-X.
138. Sekimoto, S., Nakagawa, H., Okazaki, S., Fukuda, K., Asakura, S., Shigemori, T., Takahashi, S. 2000. A fiber-optic evanescent-wave hydrogen gas sensor using palladium-supported tungsten oxide. *Sensors and Actuators B: Chemical*, *66* (1), 142–145. https://doi.org/10.1016/S0925-4005(00)00330-0.
139. Villatoro, J., Díez, A., Cruz, J. L., Andrés, M. V. 2001. Highly sensitive optical hydrogen sensor using circular Pd-coated singlemode tapered fibre. *Electronics Letters*, *37* (16), 1011–1012. https://doi.org/10.1049/el:20010716.
140. Yang, M., Sun, Y., Zhang, D., Jiang, D. 2010. Using Pd/WO_3 composite thin films as sensing materials for optical fiber hydrogen sensors. *Sensors and Actuators B: Chemical*, *143* (2), 750–753. https://doi.org/10.1016/j.snb.2009.10.017.
141. Alkhabet, M. M., Girei, S. H., Paiman, S., Arsad, N., Mahdi, M. A., Yaacob, M. H. 2020. Highly sensitive hydrogen sensor based on palladium-coated tapered optical fiber at room temperature. *Engineering Proceedings*, *2*, 8. https://doi.org/10.3390/ecsa-7-08186.

142. Butler, M. A. 1991. Fiber optic sensor for hydrogen concentrations near the explosive limit. *Journal of the Electrochemical Society*, *138* (9), L46. https://doi.org/10.1149/1.2086073.
143. Liu, Y., Chen, Y., Song, H., Zhang, G. 2012. Modeling analysis and experimental study on the optical fiber hydrogen sensor based on Pd-Y alloy thin film. *Review of Scientific Instruments*, *83* (7), 075001. https://doi.org/10.1063/1.4731725.
144. Liu, H., Nosheen, F., Wang, X. 2015. Noble metal alloy complex nanostructures: Controllable synthesis and their electrochemical property. *Chemical Society Reviews*, *44* (10), 3056–3078. https://doi.org/10.1039/C4CS00478G.
145. Bévenot, X., Trouillet, A., Veillas, C., Gagnaire, H., Clément, M. 2000. Hydrogen leak detection using an optical fibre sensor for aerospace applications. *Sensors and Actuators B: Chemical*, *67* (1), 57–67. https://doi.org/10.1016/S0925-4005(00)00407-X.
146. Mak, T., Westerwaal, R., Slaman, M., Schreuders, H., Vugt, A., Victoria, M., Boelsma, C., Dam, B. 2014. Optical fiber sensor for the continuous monitoring of hydrogen in oil. *Sensors and Actuators B Chemical*, *190*, 982–989. https://doi.org/10.1016/j.snb.2013.09.080.
147. Park, K. S., Kim, Y. H., Eom, J. B., Park, S. J., Park, M.-S., Jang, J.-H., Lee, B. H. 2011. Compact and multiplexible hydrogen gas sensor assisted by self-referencing technique. *Optics Express*, *19* (19), 18190–18198. https://doi.org/10.1364/OE.19.018190.
148. Tang, S., Zhang, B., Li, Z., Dai, J., Wang, G., Yang, M. 2015. Self-compensated microstructure fiber optic sensor to detect high hydrogen concentration. *Optics Express*, *23* (17), 22826–22835. https://doi.org/10.1364/OE.23.022826.
149. Hosoki, A., Nishiyama, M., Igawa, H., Choi, Y., Watanbea, K. 2012. Surface plasmon resonance hydrogen sensor based on hetero-core optical fiber structure. *Procedia Engineering*, *47*, 128–131. https://doi.org/10.1016/j.proeng.2012.09.101.
150. Mishra, S. K., Gupta, B. D. 2012. Surface plasmon resonance-based fiber-optic hydrogen gas sensor utilizing indium–tin oxide (ITO) thin films. *Plasmonics*, *7* (4), 627–632. https://doi.org/10.1007/s11468-012-9351-7.
151. Hosoki, A., Nishiyama, M., Igawa, H., Seki, A., Choi, Y., Watanabe, K. 2013. A surface plasmon resonance hydrogen sensor using Au/Ta_2O_5/Pd multi-layers on hetero-core optical fiber structures. *Sensors and Actuators B: Chemical*, *185*, 53–58. https://doi.org/10.1016/j.snb.2013.04.072.
152. Perrotton, C., Westerwaal, R. J., Javahiraly, N., Slaman, M., Schreuders, H., Dam, B., Meyrueis, P. 2013. A reliable, sensitive and fast optical fiber hydrogen sensor based on surface plasmon resonance. *Optics Express*, *21* (1), 382–390. https://doi.org/10.1364/OE.21.000382.
153. Wang, X., Tang, Y., Zhou, C., Liao, B. 2013. Design and optimization of the optical fiber surface plasmon resonance hydrogen sensor based on wavelength modulation. *Optics Communications*, *298–299*, 88–94. https://doi.org/10.1016/j.optcom.2013.01.054.
154. Tabassum, R., Gupta, B. D. 2015. Fiber optic hydrogen gas sensor utilizing surface plasmon resonance and native defects of zinc oxide by palladium. *Journal of Optics*, *18* (1), 015004. https://doi.org/10.1088/2040-8978/18/1/015004.
155. Takahashi, K., Hosoki, A., Nishiyama, M., Igawa, H., Watanabe, K. 2016. Long-term measurements of SPR hydrogen sensor based on hetero-core optical fiber with Au/Ta_2O_5/Pd/Au multilayers. In *Photonic Instrumentation Engineering III*, International Society for Optics and Photonics, Vol. 9754, p 975415. https://doi.org/10.1117/12.2210826.
156. Liu, H., Wang, M., Wang, Q., Li, H., Ding, Y., Zhu, C. 2018. Simultaneous measurement of hydrogen and methane based on PCF-SPR structure with compound film-coated side-holes. *Optical Fiber Technology*, *45*, 1–7. https://doi.org/10.1016/j.yofte.2018.05.007.

157. Hosoki, A., Nishiyama, M., Sakurai, N., Igawa, H., Watanabe, K. 2020. Long-term hydrogen detection using a hetero-core optical fiber structure featuring Au/Ta_2O_5/Pd/Pt multilayer films. *IEEE Sensors Journal*, *20* (1), 227–233. https://doi.org/10.1109/JSEN.2019.2942599.
158. Deng, X., Li, Z., Zhou, C., Lv, Z., Kang, C. 2019. Design and optimization of sensing layer for terminal reflective SPR optical fiber hydrogen sensor. *IOP Conference Series: Materials Science and Engineering*, *592*, 012074. https://doi.org/10.1088/1757-899X/592/1/012074.
159. Mishra, S. K., Gupta, B. D. 2012. Surface plasmon resonance-based fiber-optic hydrogen gas sensor utilizing indium–tin oxide (ITO) thin films. *Plasmonics*, *7* (4), 627–632. https://doi.org/10.1007/s11468-012-9351-7.
160. Perrotton, C., Westerwaal, R. J., Javahiraly, N., Slaman, M., Schreuders, H., Dam, B. Meyrueis, P. 2013. A reliable, sensitive and fast optical fiber hydrogen sensor based on surface plasmon resonance. *Optics Express*, *21* (1), 382–390. https://doi.org/10.1364/OE.21.000382.
161. Wang, X., Tang, Y., Zhou, C., Liao, B. 2013. Design and optimization of the optical fiber surface plasmon resonance hydrogen sensor based on wavelength modulation. *Optics Communications*, *298–299*, 88–94. https://doi.org/10.1016/j.optcom.2013.01.054.
162. Wang, Y., Yang, M., Zhang, G., Dai, J., Zhang, Y., Zhuang, Z., Hu, W. 2015. Fiber optic hydrogen sensor based on Fabry–Perot interferometer coated with sol-gel Pt/WO_3 coating. *Journal of Lightwave Technology*, *33* (12), 2530–2534. https://doi.org/10.1109/JLT.2014.2365183.
163. Takahashi, K., Hosoki, A., Nishiyama, M., Igawa, H., Watanabe, K. 2016. Long-term measurements of SPR hydrogen sensor based on hetero-core optical fiber with Au/Ta_2O_5/Pd/Au multilayers. In *Photonic Instrumentation Engineering III*, International Society for Optics and Photonics, Vol. 9754, p 975415. https://doi.org/10.1117/12.2210826.
164. Zhou, X., Ma, F., Ling, H., Yu, B., Peng, W., Yu, Q. 2020. A compact hydrogen sensor based on the fiber-optic Fabry-Perot interferometer. *Optics & Laser Technology*, *124*, 105995. https://doi.org/10.1016/j.optlastec.2019.105995.
165. Butler, M. A. 1984. Optical fiber hydrogen sensor. *Applied Physics Letters*, *45* (10), 1007–1009. https://doi.org/10.1063/1.95060.
166. Zhao, Y., Li, X., Cai, L., Yang, Y. 2015. Refractive index sensing based on photonic crystal fiber interferometer structure with up-tapered joints. *Sensors and Actuators B: Chemical*, *221*, 406–410. https://doi.org/10.1016/j.snb.2015.06.148.
167. Gong, J., Shen, C., Sun, Z., Shuai, S., Lang, T., Xiao, Y. 2018. An optical fiber hydrogen concentration sensor based on Mach-Zehnder interferometer coated with a film of palladium. In *2018 Asia Communications and Photonics Conference (ACP)*, pp 1–3. https://doi.org/10.1109/ACP.2018.8595983.
168. Kim, Y. H., Kim, M. J., Rho, B. S., Park, M., Jang, J., Lee, B. H. 2011. Ultra sensitive fiber-optic hydrogen sensor based on high order cladding mode. *IEEE Sensors Journal*, *11* (6), 1423–1426. https://doi.org/10.1109/JSEN.2010.2092423.
169. Zhou, F., Qiu, S., Luo, W., Xu, F., Lu, Y. 2014. An all-fiber reflective hydrogen sensor based on a photonic crystal fiber in-line interferometer. *IEEE Sensors Journal*, *14* (4), 1133–1136. https://doi.org/10.1109/JSEN.2013.2293347.
170. Yu, Z., Jin, L., Chen, L., Li, J., Ran, Y., Guan, B. 2015. Microfiber Bragg grating hydrogen sensors. *IEEE Photonics Technology Letters*, *27* (24), 2575–2578. https://doi.org/10.1109/LPT.2015.2478445.
171. Gu, F., Wu, G., Zeng, H. 2014. Hybrid photon–plasmon Mach–Zehnder interferometers for highly sensitive hydrogen sensing. *Nanoscale*, *7* (3), 924–929. https://doi.org/10.1039/C4NR06642A.

172. Liu, S., Shen, C., Zhang, C., Zhao, C., Lang, T., Yu, J., Fang, J. 2019. Peanut-type fiber based hydrogen sensor coated by PDMS covered with WO_3/SiO_2. In *2019 18th International Conference on Optical Communications and Networks (ICOCN)*, pp 1–3. https://doi.org/10.1109/ICOCN.2019.8933957.
173. Wang, Y., Yang, M., Zhang, G., Dai, J., Zhang, Y., Zhuang, Z., Hu, W. 2015. Fiber optic hydrogen sensor based on Fabry–Perot interferometer coated with sol-gel Pt/WO_3 coating. *Journal of Lightwave Technology*, *33* (12), 2530–2534. https://doi.org/10.1109/JLT.2014.2365183.
174. Yu, C., Liu, L., Chen, X., Liu, Q., Gong, Y. 2015. Fiber-optic Fabry-Perot hydrogen sensor coated with Pd-Y film. *Photonic Sensors*, *5*, 142–145. https://doi.org/10.1007/s13320-015-0237-0.
175. Li, Y., Zhao, C., Xu, B., Wang, D., Yang, M. 2018. Optical cascaded Fabry–Perot interferometer hydrogen sensor based on Vernier effect. *Optics Communications*, *414*, 166–171. https://doi.org/10.1016/j.optcom.2017.12.012.
176. Zhou, X., Ma, F., Ling, H., Yu, B., Peng, W., Yu, Q. 2020. A compact hydrogen sensor based on the fiber-optic Fabry-Perot interferometer. *Optics & Laser Technology*, *124*, 105995. https://doi.org/10.1016/j.optlastec.2019.105995.
177. Cao, R., Yang, Y., Wang, M., Yi, X., Wu, J., Huang, S., Chen, K. P. 2020. Multiplexable intrinsic Fabry–Perot interferometric fiber sensors for multipoint hydrogen gas monitoring. *Optics Letters*, *45* (11), 3163–3166. https://doi.org/10.1364/OL.389433.
178. Ma, J., Zhou, Y., Bai, X., Chen, K., Guan, B.-O. 2019. High-sensitivity and fast-response fiber-tip Fabry–Pérot hydrogen sensor with suspended palladium-decorated graphene. *Nanoscale*, *11* (34), 15821–15827. https://doi.org/10.1039/C9NR04274A.
179. Xu, B., Zhao, F. P., Wang, D. N., Zhao, C.-L., Li, J., Yang, M., Duan, L. 2020. Tip hydrogen sensor based on liquid-filled in-fiber Fabry–Pérot interferometer with Pt-loaded WO_3 coating. *Measurement, Science and Technology*, *31* (12), 125107. https://doi.org/10.1088/1361-6501/ab7e6a.
180. Kim, Y., Noh, T. K., Lee, Y. W., Kim, E.-S., Shin, B.-S., Lee, S.-M. 2013. Fiber-optic hydrogen sensor based on polarization-diversity loop interferometer. *Journal of the Korean Physical Society*, *62* (4), 575–580. https://doi.org/10.3938/jkps.62.575.
181. Yang, Y., Yang, F., Wang, H., Yang, W., Jin, W. 2015. Temperature-insensitive hydrogen sensor with polarization-maintaining photonic crystal fiber-based Sagnac interferometer. *Journal of Lightwave Technology*, *33* (12), 2566–2571. https://doi.org/10.1109/JLT.2014.2375362.
182. Xu, B., Zhao, C. L., Yang, F., Gong, H., Wang, D. N., Dai, J., Yang, M. 2016. Sagnac interferometer hydrogen sensor based on panda fiber with Pt-loaded WO_3/SiO_2 coating. *Optics Letters*, *41* (7), 1594–1597. https://doi.org/10.1364/OL.41.001594.
183. Darmadi, I., Nugroho, F. A. A., Langhammer, C. 2020. High-performance nanostructured palladium-based hydrogen sensors—current limitations and strategies for their mitigation. *ACS Sensors*, *5* (11), 3306–3327. https://doi.org/10.1021/acssensors.0c02019.
184. Wu, B., Zhao, C., Xu, B., Li, Y. 2018. Optical fiber hydrogen sensor with single Sagnac interferometer loop based on Vernier effect. *Sensors and Actuators B: Chemical*, *255*, 3011–3016. https://doi.org/10.1016/j.snb.2017.09.124.
185. Kim, Y. H., Kim, M. J., Rho, B. S., Kwack, K. H., Lee, B. H. 2010. Mach-Zehnder interferometric hydrogen sensor based on a single mode fiber having core structure modification at two sections. In *2010 IEEE SENSORS*, pp 1483–1486. https://doi.org/10.1109/ICSENS.2010.5690361.
186. Zhou, F., Qiu, S., Luo, W., Xu, F., Lu, Y. 2014. An all-fiber reflective hydrogen sensor based on a photonic crystal fiber in-line interferometer. *IEEE Sensors Journal*, *14* (4), 1133–1136. https://doi.org/10.1109/JSEN.2013.2293347.

187. Hu, T. Y., Wang, D. N., Wang, M., Li, Z., Yang, M. 2014. Miniature hydrogen sensor based on fiber inner cavity and Pt-doped WO_3 coating. *IEEE Photonics Technology Letters*, *26* (14), 1458–1461. https://doi.org/10.1109/LPT.2014.2327013.
188. Yu, Z., Jin, L., Sun, L., Li, J., Ran, Y., Guan, B. 2016. Highly sensitive fiber taper interferometric hydrogen sensors. *IEEE Photonics Journal*, *8* (1), 1–9. https://doi.org/10.1109/JPHOT.2015.2507369.
189. Zhao, Y., Wu, Q., Zhang, Y. 2017. High-sensitive hydrogen sensor based on photonic crystal fiber model interferometer. *IEEE Transactions on Instrumentation and Measurement*, *66* (8), 2198–2203. https://doi.org/10.1109/TIM.2017.2676141.
190. Zhang, C., Shen, C., Liu, S., Fang, J., Sun, Z., Gong, J., Ding, Z., Lang, T., Zhao, C., Chen, H. 2019. Cascaded TCF with WO_3 film based Mach–Zehnder interferometer for hydrogen sensing. In *2019 18th International Conference on Optical Communications and Networks (ICOCN)*, pp 1–3. https://doi.org/10.1109/ICOCN.2019.8934171.
191. Wang, M., Yang, M., Cheng, J., Zhang, G., Liao, C. R., Wang, D. N. 2013. Fabry–Pérot interferometer sensor fabricated by femtosecond laser for hydrogen sensing. *IEEE Photonics Technology Letters*, *25* (8), 713–716. https://doi.org/10.1109/LPT.2013.2241421.
192. Wang, Y., Yang, M., Zhang, G., Dai, J., Zhang, Y., Zhuang, Z. 2014. Ultra-highly sensitive hydrogen sensor based on fiber Fabry-Perot interferometer with Pt/WO_3 coating. In *23rd International Conference on Optical Fibre Sensors*, International Society for Optics and Photonics, Vol. 9157, p 91574F. https://doi.org/10.1117/12.2058845.
193. Zhang, G., Yang, M., Wang, Y. 2014. Optical fiber-tip Fabry–Perot interferometer for hydrogen sensing. *Optics Communications*, *329*, 34–37. https://doi.org/10.1016/j.optcom.2014.04.084.
194. Yu, C., Liu, L., Chen, X., Liu, Q., Gong, Y. 2015. Fiber-optic Fabry-Perot hydrogen sensor coated with Pd-Y film. *Photonic Sensors*, *5* (2), 142–145.
195. Xu, B., Li, P., Wang, D. N., Zhao, C.-L., Dai, J., Yang, M. 2017. Hydrogen sensor based on polymer-filled hollow core fiber with Pt-loaded WO_3/SiO_2 coating. *Sensors and Actuators B: Chemical*, *245*, 516–523. https://doi.org/10.1016/j.snb.2017.01.206.
196. Shao, J., Xie, W., Song, X., Zhang, Y. 2017. A new hydrogen sensor based on SNS fiber interferometer with Pd/WO_3 coating. *Sensors*, *17* (9), 2144. https://doi.org/10.3390/s17092144.
197. Li, Y., Zhao, C., Xu, B., Wang, D., Yang, M. 2018. Optical cascaded Fabry–Perot interferometer hydrogen sensor based on Vernier effect. *Optics Communications*, *414*, 166–171. https://doi.org/10.1016/j.optcom.2017.12.012.
198. Xu, B., Zhao, F. P., Wang, D., Zhao, C.-L., Li, J., Yang, M., Duan, L. 2020. Tip hydrogen sensor based on liquid filled in-fiber Fabry-Perot interferometer with Pt-loaded WO_3 coating. *Measurement Science and Technology*, 31 (12), 125107. https://doi.org/10.1088/1361-6501/ab7e6a.
199. Xu, B., Zhao, C. L., Yang, F., Gong, H., Wang, D. N., Dai, J., Yang, M. 2016. Sagnac interferometer hydrogen sensor based on panda fiber with Pt-loaded WO_3/SiO_2 coating. *Optics Letters*, *41* (7), 1594–1597. https://doi.org/10.1364/OL.41.001594.
200. Grobnic, D., Mihailov, S. J., Walker, R. B., Cuglietta, G., Smelser, C. W. 2011. Hydrogen detection in high pressure gas mixtures using a twin hole fibre Bragg grating. In Proc*eedings* SPIE 7753, 21st International Conference on Optical Fiber Sensors, p 77537D. https://doi.org/10.1117/12.885954.
201. Zalvidea, D., Diez, A., Cruz, J. L., Andres, M. V. 2004. Wavelength multiplexed hydrogen sensor based on palladium-coated fibre-taper and Bragg grating. *Electronics Letters*, *40* (5), 301–302. https://doi.org/10.1049/el:20040211.

202. Yang, S., Dai, J., Qin, Y., Xiang, F., Wang, G., Yang, M. 2018. Improved performance of fiber optic hydrogen sensor based on MoO_3 by ion intercalation. *Sensors and Actuators B: Chemical*, *270*, 333–340. https://doi.org/10.1016/j.snb.2018.05.060.
203. Massaroni, C., Caponero, M. A., D'Amato, R., Lo Presti, D., Schena, E. 2017. Fiber Bragg grating measuring system for simultaneous monitoring of temperature and humidity in mechanical ventilation. *Sensors*, *17* (4), 749. https://doi.org/10.3390/s17040749.
204. Trouillet, A., Marin, E., Veillas, C. 2006. Fibre gratings for hydrogen sensing. *Measurement Science and Technology*, *17* (5), 1124–1128. https://doi.org/10.1088/0957-0233/17/5/S31.
205. Buric, M., Chen, K. P., Bhattarai, M., Swinehart, P. R., Maklad, M. 2007. Active fiber Bragg grating hydrogen sensors for all-temperature operation. *IEEE Photonics Technology Letters*, *19* (5), 255–257. https://doi.org/10.1109/LPT.2006.888973.
206. Ma, G., Li, C., Mu, R., Jiang, J., Luo, Y. 2014. Fiber Bragg grating sensor for hydrogen detection in power transformers. *IEEE Transactions on Dielectrics and Electrical Insulation*, *21* (1), 380–385. https://doi.org/10.1109/TDEI.2013.004381.
207. Samsudin, M. R., Shee, Y. G., Adikan, F. R. M., Razak, B. B. A., Dahari, M. 2016. Fiber Bragg gratings hydrogen sensor for monitoring the degradation of transformer oil. *IEEE Sensors Journal*, *16* (9), 2993–2999. https://doi.org/10.1109/JSEN.2016.2517214.
208. Caucheteur, C., Debliquy, M., Lahem, D., Megret, P. 2008. Catalytic fiber Bragg grating sensor for hydrogen leak detection in air. *IEEE Photonics Technology Letters*, *20* (2), 96–98. https://doi.org/10.1109/LPT.2007.912557.
209. Yang, M., Yang, Z., Dai, J., Zhang, D. 2012. Fiber optic hydrogen sensors with sol–gel WO_3 coatings. *Sensors and Actuators B: Chemical*, *166–167*, 632–636. https://doi.org/10.1016/j.snb.2012.03.026.
210. Sutapun, B., Tabib-Azar, M., Kazemi, A. 1999. Pd-coated elastooptic fiber optic Bragg grating sensors for multiplexed hydrogen sensing. *Sensors and Actuators B: Chemical*, *60* (1), 27–34. https://doi.org/10.1016/S0925-4005(99)00240-3.
211. Aleixandre, M., Corredera, P., Hernanz, M. L., Sayago, I., Horrillo, M. C., Gutierrez-Monreal, J. 2007. Study of a palladium coated Bragg grating sensor to detect and measure low hydrogen concentrations. In *2007 Spanish Conference on Electron Devices*, pp 223–225. https://doi.org/10.1109/SCED.2007.384032.
212. Guoming, M., Li, C., Luo, Y., Mu, R., Wang, L. 2012. High sensitive and reliable fiber Bragg grating hydrogen sensor for fault detection of power transformer. *Sensors and Actuators B: Chemical*, 169, 195–198. https://doi.org/10.1016/j.snb.2012.04.066.
213. Yang, M., Yang, Z., Dai, J., Zhang, D. Fiber optic hydrogen sensors with sol–gel WO_3 coatings. 2012. *Sensors and Actuators B: Chemical*, *166–167*, 632–636. https://doi.org/10.1016/j.snb.2012.03.026.
214. Dai, J., Yang, M., Yu, X., Lu, H. 2013. Optical hydrogen sensor based on etched fiber Bragg grating sputtered with Pd/Ag composite film. *Optical Fiber Technology*, *19* (1), 26–30. https://doi.org/10.1016/j.yofte.2012.09.006.
215. Dai, J., Yang, M., Yang, Z., Li, Z., Wang, Y., Wang, G., Zhang, Y., Zhuang, Z. 2014. Enhanced sensitivity of fiber Bragg grating hydrogen sensor using flexible substrate. *Sensors and Actuators B: Chemical*, *196*, 604–609. https://doi.org/10.1016/j.snb.2014.02.069.
216. Ma, G., Li, C., Mu, R., Jiang, J., Luo, Y. 2014. Fiber Bragg grating sensor for hydrogen detection in power transformers. *IEEE Transactions on Dielectrics and Electrical Insulation*, *21* (1), 380–385. https://doi.org/10.1109/TDEI.2013.004381.

217. Coelho, L., de Almeida, J. M. M. M., Santos, J. L., Viegas, D. 2015. Fiber optic hydrogen sensor based on an etched Bragg grating coated with palladium. *Applied Optics*, *54* (35), 10342–10348. https://doi.org/10.1364/AO.54.010342.
218. Jiang, J., Ma, G., Li, C., Song, H., Luo, Y., Wang, H. 2015. Highly sensitive dissolved hydrogen sensor based on side-polished fiber Bragg grating. *IEEE Photonics Technology Letters*, *27* (13), 1453–1456. https://doi.org/10.1109/LPT.2015.2425894.
219. Masuzawa, S., Okazaki, S., Maru, Y., Mizutani, T. 2015. Catalyst-type-an optical fiber sensor for hydrogen leakage based on fiber Bragg gratings. *Sensors and Actuators B: Chemical*, *217*, 151–157. https://doi.org/10.1016/j.snb.2014.10.026.
220. Zhong, X., Yang, M., Huang, C., Wang, G., Dai, J., Bai, W. 2016. Water photolysis effect on the long-term stability of a fiber optic hydrogen sensor with Pt/WO_3. *Scientific Reports*, *6* (1), 39160. https://doi.org/10.1038/srep39160.
221. Dai, J., Peng, W., Wang, G., Xiang, F., Qin, Y., Wang, M., Dai, Y., Yang, M., Deng, H., Zhang, P. 2017. Ultra-high sensitive optical fiber hydrogen sensor using self-referenced demodulation method and WO_3-Pd_2Pt-Pt composite film. *Optics Express*, *25* (3), 2009–2015. https://doi.org/10.1364/OE.25.002009.
222. Zhou, X., Dai, Y., Karanja, J. M., Liu, F., Yang, M. 2017. Microstructured FBG hydrogen sensor based on Pt-loaded WO_3. *Optics Express*, *25* (8), 8777–8786. https://doi.org/10.1364/OE.25.008777.
223. Hunze, A., Badcock, R. A., Fisser, M. 2018. Response time of a fiber Bragg grating based hydrogen sensor for transformer monitoring. *Proceedings*, *2* (13), 745. https://doi.org/10.3390/proceedings2130745.
224. Yu, J., Wu, Z., Yang, X., Han, X., Zhao, M. 2018. Tilted fiber Bragg grating sensor using chemical plating of a palladium membrane for the detection of hydrogen leakage. *Sensors*, *18* (12), 4478. https://doi.org/10.3390/s18124478.
225. Cai, S., Liu, F., Wang, R., Xiao, Y., Li, K., Caucheteur, C., Guo, T. 2020. Narrow bandwidth fiber-optic spectral combs for renewable hydrogen detection. *Science China Information Sciences*, *63* (12), 222401. https://doi.org/10.1007/s11432-020-3058-2.
226. Zhou, X., Dai, Y., Karanja, J. M., Liu, F., Yang, M. 2017. Microstructured FBG hydrogen sensor based on Pt-loaded WO_3. *Optics Express*, *25* (8), 8777–8786. https://doi.org/10.1364/OE.25.008777.
227. Dai, J., Yang, M., Yu, X., Lu, H. 2013. Optical hydrogen sensor based on etched fiber Bragg grating sputtered with Pd/Ag composite film. *Optical Fiber Technology*, *19* (1), 26–30. https://doi.org/10.1016/j.yofte.2012.09.006.
228. Dai, J., Yang, M., Yang, Z., Li, Z., Wang, Y., Wang, G., Zhang, Y., Zhuang, Z. 2014. Enhanced sensitivity of fiber Bragg grating hydrogen sensor using flexible substrate. *Sensors and Actuators B: Chemical*, *196*, 604–609. https://doi.org/10.1016/j.snb.2014.02.069.
229. Dai, J., Peng, W., Wang, G., Xiang, F., Qin, Y., Wang, M., Yang, M., Deng, H., Zhang, P. 2017. Improved performance of fiber optic hydrogen sensor based on WO_3-Pd_2Pt-Pt composite film and self-referenced demodulation method. *Sensors and Actuators B: Chemical*, *249*, 210–216. https://doi.org/10.1016/j.snb.2017.04.103.
230. Fisser, M., Badcock, R. A., Teal, P. D., Janssens, S., Hunze, A. 2018. Palladium-based hydrogen sensors using fiber Bragg gratings. *Journal of Lightwave Technology*, *36* (4), 850–856. https://doi.org/10.1109/JLT.2017.2713038.
231. Zhang, G., Yang, M., Wang, M., Li, D., Wang, X. 2013. Refractometer based on a microslot in single-multi-single fiber fabricated by femtosecond laser. *Optical Engineering*, *52* (4), 044401. https://doi.org/10.1117/1.OE.52.4.044401.

232. Wen, C., Huang, X., Shen, C. 2020. Multiple-pass enhanced Raman spectroscopy for fast industrial trace gas detection and process control. *Journal of Raman Spectroscopy, 51* (5), 781–787. https://doi.org/10.1002/jrs.5838.
233. Coelho, L., de Almeida, J. M. M. M. Santos, J. L., Viegas, D. 2015. Fiber optic hydrogen sensor based on an etched Bragg grating coated with palladium. *Applied Optics, 54* (35), 10342–10348. https://doi.org/10.1364/AO.54.010342.

9 Future Promise and Issues in Flexible Transparent Electronics and Optoelectronics

Jyoti Mali and Om Prakash

CONTENTS

9.1 INTRODUCTION

Transparent electronics have recently attracted numerous applications in flexible and recyclable electronics and optoelectronics due to their strong potential and simple and cost-effective fabrication process. The traditional component is transparent electrode indium tin oxide (ITO); though this has numerous applications in organic electronics, it is brittle in nature and has a significantly high cost, leading to limitations for application in next-generation devices, specifically flexible electronics [24–26]. Graphene, carbon nanotubes (CNTs), metal nanowires, different conductive polymers, and metallic nanofibers are the alternatives at the forefront of flexible transparent electrodes (FTEs) [1]. Among these, silver nanowire (AgNW) network has outstanding electrical conductivity, high transmittance, low cost, and easy availability. FTEs fabricated with AgNW are a promising alternative to ITO films in optoelectronics device applications [27, 28]. Despite the fact that CNTs have great electrical, mechanical, and thermal properties, CNT electrodes show lower electrical conductivity compared to ITO electrodes because of their enormous contact oppositions and broad packaging of the CNTs. Graphene has higher Fermi velocity, of approximately 106 m/s, and intrinsic in-plane conductivity [25]. However, mass production of the higher-performance graphene film remains a significant issue. Even though the fabrication process using chemical vapor deposition has potential to produce larger-area, higher-performance graphene, the process costs a great deal and also requires extremely high temperatures.

The next-generation devices or cutting-edge gadgets open up a wide scope for new applications and innovations; for example, solar cells, energy-harvesting devices and sensors, flexible lights, display technologies in the architecture, consumer electronics, and textile industries, biomedical applications, automobile industry applications, robotics, and defense applications [29–32]. As conventional electronic components that are highly capable of all these applications, it is expected that the mechanical features of FTEs will be explored by adhering to the novel form factors throughout the hybrid strategies, or to independent applications which don't require higher computational power and are predetermined to be highly robust for deformation, have lower cost, and are thin and disposable in nature [2, 3].

Metallic nanowire-based electrodes are the most promising alternatives to ITO with its predominant optical, mechanical, and electrical properties [15–16]. Both irregular and ordinary metallic nanowire networks have received growing interest from researchers, academics, and industry. The random metallic nanowires can be scattered in solvent and stored on substrates through ease arrangement-based preparation. This makes the nanowire-based electrodes more compatible for higher-throughput applications and larger-area production of next-generation FTEs and flexible optoelectronic devices. Additionally, for normal metallic nanowire-based electrodes, called metal lattice, electrical conductivity and optical visibility can be effortlessly improved by changing the geometric parameters of the structure of the nanowires [17–23]. At the point when the metal lattice is in sub-micrometer scale and the line width is near sub-wavelength, metal cross-sections can be considered appropriate mass materials to assess the sheet obstruction of the films. Different metallic materials (e.g. gold, silver, and copper) are used to accomplish diverse work capacities and substance properties for different applications. Silver, a material with high electrical conductivity and somewhat low cost, is considered the most appropriate nanowire material. Furthermore, the general execution of AgNW terminals has just outperformed that of ITO electrodes.

A comparison between ITO, CNTs, AgNW, and graphene is shown in Table 9.1. It was reported that the application of graphene electrodes in solar cells and organic electronics yields power conversion efficiency from 6.1% to 7.1 % [16, 17]. Here the comparison parameters consist of transmittance (T) and sheet resistance (R_S), where T is calculated as $T = (1 + [Z_0/2\ R_S]\ [\sigma_0/\sigma_2D])^{-2}$. Hence the transparency is directly related to the sheet resistance (R_S), where the free space impedance is Z_0, and optical and 2D conductivity are σ_0 and σ_2D, respectively. The sheet resistance (R_S) is proportional to the number of layers (N) by $R_S = (\sigma_2DN)^{-1}$ [26, 36].The other comparison parameters shown in Table 9.1 are stability, haze, and conductivity, from which we can conclude that materials such as CNTs, AgNWs, and graphene have more promising applications and features that can, with few challenges and limitations, make next-generation optoelectronic devices lighter in weight and more cost-effective and eco-friendly compared to traditional ITOs.

TABLE 9.1
Comparison of different properties of transparent flexible electronics and optoelectronics devices

Material	Flexibility	Stability	Conductivity	Transmittance (T%)	Sheet Resistance Rs (Ω/sq)	Haze	Reference
Indium tin oxide	Inferior	Average	Good	90	18,537	Low	[40]
Carbon nanotubes	Good	Good	Average	82–88	<300	High	[41, 42]
Graphene	Good	Excellent	Average	>90	100–400	Low	[47, 48]
Silver nanowires	Average	Average	Good	90	30–50	High	[43]

In this chapter, we focus on ongoing research and advances in the fabrication process of flexible transparent AgNW electrodes. Initially we present a few basic necessities with respect to the electrical, optical, thermal, and mechanical properties of flexible transparent AgNW electrode films in different applications [21, 22]. Second, the synthesis of AgNW and the film-shaping procedures for adaptable, transparent AgNW electrodes is discussed. Third, the ongoing evaluations of streamlining the properties of FTEs will be described in detail. Finally, future difficulties in the inescapable appropriation of flexible transparent AgNW electrodes is demonstrated.

9.2 FABRICATION OF FLEXIBLE TRANSPARENT ELECTRODES BASED ON CARBON NANOTUBE

Transparent flexible electrodes (TFEs) are thin film transistors made of materials that are optically transparent and electrically conductive by nature. ITO has been used most commonly for transparent and conductive electrodes (TCEs) in inflexible electronics because of its outstanding electrical and optical properties.

As of late, adaptable TCEs manufactured by solution methods, which have a few favorable features (e.g. straightforward and constant cycle and moderately low expenses), have attracted attention for many applications and aims due to their wide scope for use in flexible electronics such as display devices, contact screen boards and touchscreen panels, sensors, and film warmers that can join to skin, textures, and papers [2, 3]. However, ITO has disadvantages where adaptability is required, because of its fragile nature due to low crack strain and lack of adaptability, a high preparation temperature that harms the adaptable substrates, a low bond to polymeric materials, and overall lack of availability which makes their cost variable [7]. Hence, this has as of late propelled different investigations to find elective materials to supplant ITO films, including metal lattices, AgNW, conductive polymers, graphene, and CNTs [4, 5]. Likewise, for adaptable applications, TCEs must be kept on polymer substrates such as polyethylene terephthalate (PET), polypropylene, and polydimethylsiloxane.

The unprecedented electrical and mechanical properties of CNTs make them perfect for adaptable and flexible electrodes, particularly superior adaptable and flexible integrated circuits, which are the center units of electronic frameworks for data preparation. As we can see, CNTs have great advantages in the field of transparent flexible electronics.

CNTs have supreme chemical stability and thermal and electrical conductivity which offer high intrinsic conductivity, mechanical strength, flexibility, solution process ability, and potential for production at a lower cost. Based on these advantages, CNTs have broad applications in FTEs. The research scope for improving the electrical conductivity and the transmittance of CNT TCEs has increased tremendously. However, until now, TCEs fabricated by CNTs have had limitations for certain applications of electronic devices [27]. Various hybrid types of CNT-based TCEs, however, could have potential in next-generation flexible and stretchable electronics.

Enormous progress has been made in single-wire carbon nanotube (SWCNT) flexible and stretchable electronics devices. Even so, practically no SWCNT adaptable electronic item is financially accessible right now [6]. A few difficulties have to be overcome before SWCNT electronic gadgets and frameworks can be prepared for applications in commercial markets. In terms of materials, in spite of the fact that semiconductor-enhanced SWCNTs are now economically accessible in large volumes, there is still enormous lack of homogeneity regarding chirality and nanotube length. High levels of perfection and homogeneity in the base material is beneficial for uniform device execution. Also, longer nanotubes are preferred to decrease the quantity of cylinder-to-tube intersections, which could prompt further improvement in gadget portability. Be that as it may, the disintegration and detachment of long nanotubes (>10 μm) are difficult. Moreover, the impacts of surfactants on gadget electrical qualities require more exhaustive examination. The surfactants used to scatter SWCNTs are hard to eliminate and can hinder electronic conduction and, hence, increase contact and channel opposition. In the future, new small angle neutron scattering (SANS) techniques to viably break down SWCNTs without harming or shortening them should be investigated. Ongoing investigations of scattering SWCNTs utilizing super acids or salt–alkali arrangements are promising [37, 38]. Different issues facing scientists include strategies to acquire air-stable n-type conduction in SWCNTs and to improve the consistency, yield, and steadiness of SWCNT-based devices.

Future work on printed SWCNT gadgets likewise ought to zero in on improving the metal contacts and growing new dielectric materials. Stretchable and flexible devices are moderately new and have attracted significant interest. Regardless of the great stretchability of both SWCNT networks and SWCNT slender film anodes, the need for consistent dielectrics and strong interfaces has all the earmarks of becoming a bottleneck in this field. Besides, completely printed stretchable frameworks do not appear possible at present. Lastly, graphene and other 2D semiconducting materials also have expanding potential for stretchable devices [6, 8, 9, 14]. Joining SWCNTs with different types of nanomaterials may prompt some exciting outcomes [13].

CNTs have a rounded structure, flawlessly created from graphene sheets [7]. From the perspective of advancing pragmatic applications in thin film transistors (TFTs) and TCFs, it is important to integrate high-purity CNTs. Large-scale amalgamation approaches for CNTs include bend release, laser removal, and chemical vapor deposition [32–36], and numerous sorts of CNT items have been obtained, including vertically adjusted clusters, evenly adjusted exhibits, powder, and irregular dainty films. A few hundred tons of CNT has been utilized as unique additives in the batteries and composites industries [10–12]; for their utilization in TCFs and straightforward cathodes, a higher caliber of material is required.

The fabrication process of top-gated CNT TFTs on a PEN (polyethylene naphthalate (poly(ethylene 2,6-naphthalate) substrate using photosensitive dry film patter method is shown in Figure 9.1. The schematic of active channel layers with the passive elements of drain, gate, and source electrodes and the interconnections

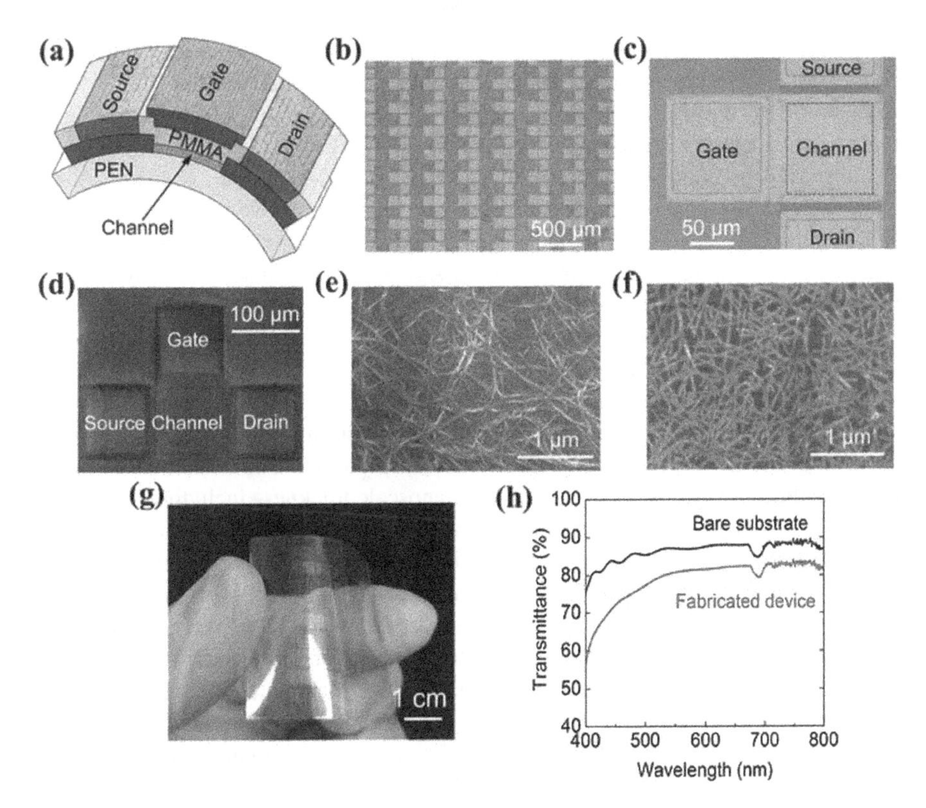

FIGURE 9.1 Flexible and transparent all-CNT TFTs. (a) Schematic of a top-gate TFT on a PEN substrate where the source, drain, and gate electrodes consist of a thick CNT film, the channel is a semiconducting CNT film, and the gate dielectric is a PMMA layer. (b) Optical microscopy image of a top-gate all-CNT TFT array. (c) Magnified image showing the structures of the source, drain, gate, and channel patterns. (d–f) SEM images of a top-gate all-CNT TFT, a CNT electrode, and a CNT channel. (g) Photograph of an all-carbon device fabricated on a flexible PEN substrate. (h) Optical transmittance of a bare substrate (top line) and the device fabricated on the substrate (bottom line).

Source: Copyright CC BY 4.0 [43]

which are made up of thin film CNT with dielectric layers of 600-nm polymethyl methacrylate (PMMA) are shown in Figure 9.1 (a). The optical microscopy images of the TFTs and their arrays are shown in Figure 9.1 (b) and (c). The channel length (L) of both the TFTs is 100 μm [43]. Figure 9.1 (d–f) shows the different SEM images for the TFTs, the CNT electrodes, and the CNT channels, respectively. Figure 9.1 (g) and (h) demonstrate the flexibility and transparency of the device. The transmittance (T) of the CNT TFT device range is ≈ 81%, exactly 10% lower than the bare PEN [43].

9.3 FABRICATION OF FLEXIBLE TRANSPARENT ELECTRODES BASED ON SILVER NANOWIRES

The fabrication process using AgNWs for transparent flexible electrodes is discussed here. Basically there are two main techniques for fabricating the TFEs using AgNW: spray coating and atmospheric pressure spatial atomic layer deposition. In this process, the transparency and electrical conductivity of the AgNW nanoelectrodes are improved by controlling the density of material. The thermal, mechanical, and electrical stability of the nanoelectrodes are drastically improved compared to those of conventional AgNWs. Here the physical model is based on the non-negligible donatives of the percolating clusters of the AgNWs for its material densities below the threshold values. The obtained results provide a means to understand and predict the physical characteristics of such nanoelectrodes for a variety of applications in electronics and optoelectronics devices.

The transparent and conductive nanocomposite with AgNW electrode arrays for applications such as skin-attachable loudspeakers and microphones is demonstrated by Wang et al. [42]. A very ultra-thin, transparent, and conductive hybrid nanomembrane (NM), within the range of nanoscale thickness including the orthogonal AgNW electrode arrays implanted in the polymer matrix, is demonstrated in Figure 9.2. Wang et al. addressed the NM-based conformal electronics requirements for a few applications, which can be further investigated for applications in wearable sensor devices, biomedical devices, and acoustic devices. The hybrid NM significantly enhances the mechanical and electrical behaviors of the ultra-thin polymer membranes [42]. A skin-attachable NM loudspeaker that can exhibit great enhancement in thermacoustic properties without any limitation in terms of heat loss from the substrate is shown in Figure 9.2. Figure 9.2 (a) shows a schematic of the fabrication procedure for the TFEs and hybrid NMs using AgNW electrodes as nanoreinforcement and a polymer matrix. Figure 9.2 (b) shows hybrid NMs suspended on the surface of the solution. Figure 9.2 (c) shows the dark field optical graphs in micro ranges for an orthogonal AgNW array produced by the bar-coating process. Figure 9.2 (d) and (e) show the cross-sectional scanning electron microscopy (SEM) image of the hybrid NMs possessed of orthogonal AgNW arrays implemented in the polymer matrix, and the optical transmittance of polymer NM, hybrid NMs, and bare PET with glass. In Figure 9.2 (f) a hybrid NM on the surface of the water image is demonstrated. A transparent and flexible hybrid NM transferred to a curved and human surface/skin is shown in Figure 9.2 (g–i).

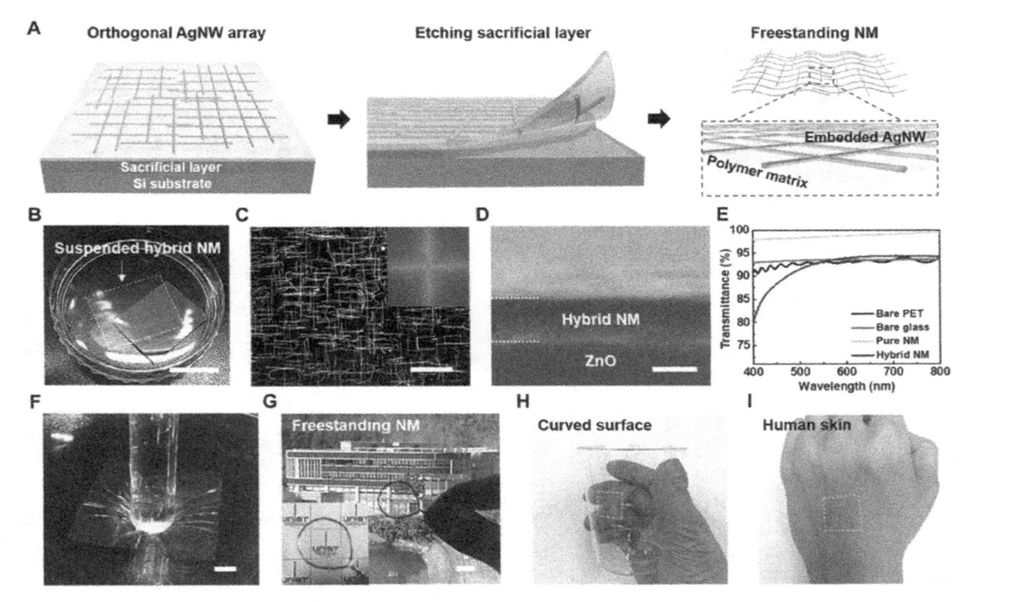

FIGURE 9.2 Fabrication of freestanding hybrid NMs with orthogonal AgNW arrays. (a) Schematic of fabrication procedure for freestanding hybrid NMs with orthogonal AgNW arrays embedded in a polymer matrix. (b) Freestanding AgNW composite NMs floating on the surface of water. Scale bar, 1 cm. (c) Dark-field optical microscope image of orthogonal AgNW arrays. The inset shows a fast Fourier transform image of the optical micrograph, corresponding to its surface geometric structure. Scale bar, 40 μm. (d) Cross-sectional SEM image of an as-fabricated hybrid NM on a ZnO/Si substrate. Scale bar, 100 nm. (e) Optical transmittance of polymer NMs, hybrid NMs, bare PET, and bare glass in the visible range of 400 nm to 800 nm. The air is used as a reference. (f) Photograph of a hybrid NM on the surface of water under compressive force applied by a glass rod. Scale bar, 3 mm. (g) A freestanding hybrid NM supported by a wire loop. The inset shows the high transparency of the hybrid NM. Scale bar, 1 cm. Hybrid NMs transferred onto (h) curvilinear surfaces and (i) human skin.

Source: Copyright 2018, Creative Commons Attribution-Non Commercial license 4.0 (CC BY-NC) [42].

9.4 FUTURE SCOPE AND APPLICATIONS OF TRANSPARENT FLEXIBLE ELECTRODES IN ELECTRONICS AND OPTICAL ELECTRONICS

Flexible transparent electronics as a growing and exciting research and development field has incredible interest for the issue of how to make adaptable electronic materials that offer both toughness and superiority at stressed states. With the appearance of on-body wearable and implantable hardware and expanding requests for human-friendly, astute, intricate robots, tremendous effort is being made to produce exceptionally adaptable materials, particularly stretchable electrodes, by both the theoretical and new networks. Among various twisting modes, stretchability is the most demanding and testing. Hence, in this chapter we tried to show the basics of the most recent advances in stretchable straightforward anodes and cathodes by comparing their characteristics and examining new advances in novel applications, including skin-like electronics, implantable biodegradable gadgets, and bioinspired intricate mechanical technology and robotics. By contrasting the optoelectrical and mechanical properties of various anode materials, their benefits and disadvantages have been observed by many research scholars and scientists to date. Key plans as far as calculations, substrates, and attachment are also discussed, offering insight into the general methodologies for designing stretchable electronics. It is proposed that profoundly stretchable and biocompatible electrodes will significantly help advance next-generation smart, life-like, automated, and flexible gadgets.

9.4.1 Why Flexible Electronics?

Adaptable hardware, namely flexible electronics and optoelectronics, are in vogue for two reasons. To start with, this guarantees a totally new, advanced design tool. Envision, for instance, small cell phones that fold over our wrists and adaptable displays with lay out as large as a TV. On the other hand, imagine photovoltaic cells and reconfigurable transmission and reception antennas that adjust to rooftops and vehicle trunks, or adaptable inserts that can screen and treat malignancy or help paraplegics walk again [10–14].

Second, adaptable hardware of flexible transparent electronic devices may cost much less to manufacture. Traditional semiconductors require complex cycles and multibillion-dollar foundries for its fabrication process. Specialists and researchers aim to print adaptable gadgets of flexible transparent electronic devices on plastic film, similar to the way we print ink on paper. As Gomez says: "In the event that we could make adaptable hardware adequately modest, you could have expandable gadgets. You could wear your telephone on your attire, or run a bioassay to evaluate your wellbeing basically by cleaning your nose with a tissue."

Before any of this occurs, however, scientists need to reconsider electronics. None of the adaptable electronic gadgets in development would facilitate the billions of semiconductors that currently fit on silicon chips or their billions of on–off cycles every second. They would not need to. In spite of tests, adaptable gadgets guarantee changes that go past collapsing displays, cheap sun-powered cells, radio wires, and sensors [38]. They could veer off in different ways; for example, assisting paraplegics to walk once more. Below are a few current issues for TFEs.

1. Advancements in slender, light, and flexible transparent electronics are behind many new developments, from curved TVs to glucose-observing contact focal points.
2. The revelation that polymers can be semiconductors has prompted the improvement of printable sun-based cells and adaptable screens utilizing natural light-producing diodes in LEDs.
3. Biocompatible electronic gadgets that can flex and stretch have scope for observing clinical variables and well-being in health-monitoring applications.

9.4.2 Applications of Transparent Flexible Electronics

Flexible transparent electronics open up various applications such as in memory and storage devices, telecom sectors, medical applications using wearable devices and bioengineering, different display panels, touch screens, solar cells, and sensors. TFEs have also opened up great scope for researchers to investigate adaptabilities in the fields of human interactivity devices, computation, and energy generation cells, as well as in electronic textiles industries. By improving the light output power, multiple applications in the field of optical, electrical, mechanical, and thermal transparent devices can be achieved [42–44]. From Figure 9.3, we can see that flexible electronics have opened up many possibilities for future applications in a variety of fields. The main characteristics of transparent flexible electronics, which make these more advanced compared to the traditional semiconductor devices and its applications, are shown in Figure 9.3.

The natural electronic materials market is ready to show remarkable development driven by expanding applications in ordinary and specialty areas, alongside growth in various applications from wide-territory presentations and radio-frequency identification labels to memories. Ease of creation and a significant level of adaptability are the two key elements driving the interest in natural hardware and its expanding infiltration into the standard market. The materials are likewise helpful for new and inventive

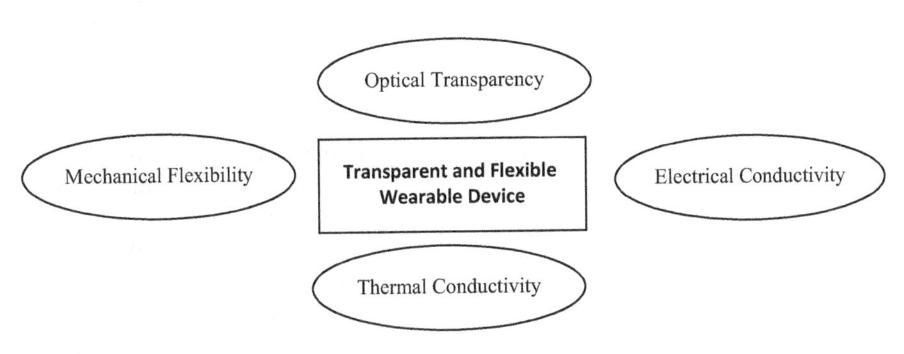

FIGURE 9.3 Characteristics of transparent flexible electronics

applications; for example, electronic paper and smart windows, wherein inorganic channels (silicon or copper) can't be utilized. Natural conductors are also expected to find application in emerging areas; for example, atomic processing.

Printed hardware, another industry that is currently undeveloped, can shape the eventual fate of printing, just as pretty much every other industry on the planet. Printed hardware is the utilization of conductive and electronic parts onto a material. This probably won't happen with a traditional fabrication process, yet the applied items will be fluid or semi-fluid in structure and will cover a substrate design.

9.5 CONCLUSION

Transparent and flexible electronics is a rising field of science and assembling innovation, which allows planting of electronic gadgets onto similar plastic substrates [16]. The area of adaptable hardware, like large-scale gadgets, natural gadgets, plastic gadgets, and printed gadgets, is driven by worldwide interest for lighter and more modest electronic items that use less energy. Because these gadgets are more resistant to strain, easy to fabricate and can flex or twist, they can be incorporated into compact gadgets, attire, and bundling materials [39].

While the market for adaptable gadgets and TFEs is as yet undeveloped, different new applications are being established by colleges, research organizations, and other cutting-edge organizations [39]. The innovation is finding expanding applications in different areas; for example, military devices, PCs, laptops, consumer items, aviation, cars and clinical devices [41]. The primary end uses for adaptable hardware include electronic presentations, photovoltaic sensors, lighting, and cognition/memory applications, among others. The capacity to construct adaptable gadgets or electronic circuits that can be extended and twisted as necessary, allows the current capacities of unbending circuits to be overtaken [38–40]. Further are enabled by the improvement of reasonable cycles that encourage the cost-efficient assembling of flexible electronics segments and the joining of unbending segments for different applications.

Emerging electrical application sectors, including warmers, solar cells, batteries, and lighting, are driving the development of the overall flexible electronics market [30]. The innovations allow electronic frameworks to be rolled, extended, washed, and worn, and subsequently adjust to different engineering highlights [32]. Hence the flexible transparent electronics, due to its adaptable gadgets segments, are likewise equipped for biodegradation with different layer structures, which decrease the states required and the number of interconnections while expanding the dependability of the framework [33].

The aim of this chapter was to describe the procedures for creation of adaptable straightforward AgNW cathodes, known as transparent flexible electronics, and to compare these to the CNT transparent flexible electronic fabrication process and the endeavors made to improve throughput. In spite of the fact that AgNW terminals allegedly show performance comparable to ITO anodes, there is as yet far to go before commercialization is possible. Right off the bat, new union strategies for tweaking the elements of AgNWs are required. AgNW cathodes have a cozy relationship with the components of silver AgNWs [16–20]. Besides, metals other than silver need examination to lessen the expense of cathodes with comparable properties; for example, copper. Cross-breed materials, such as center-shell Cu-Ni nanowires in a sandwich structure, are additionally of interest. Also, the streamlining of dependability in genuine conditions is currently deficient. The assessment of dependability is much needed to demonstrate the potential of incorporating nanowires into future gadgets. Lastly, the harmfulness of nanowires needs consideration prior to their being incorporated into business gadgets [45].

It warrants saying that a wide scope of utilizations will materialize in light of the fact that metal nanowires have great straightforward conductive execution and mechanical adaptability. Metal nanowire films permit a great deal of innovation; for example, adaptable-level display screens, electronic skins, straightforward heaters, and flexible electronics. There are still a few issues that should be settled for specific applications; for example, low surface substrates with flat properties and the natural security and achievement proportion of metal nanowires in the subsequent devices [45–46].

REFERENCES

1. Wang C, Hwang D, Yu Z, Takei K, Park J, Chen T et al. 2013. User-interactive electronic skin for instantaneous pressure visualization. Nature Materials 12(10):899–904.
2. Hammock ML, Chortos A, Tee BC-K, Tok JB-H, Bao Z. 2013. 25th anniversary article: The evolution of electronic skin (E-skin): A brief history, design considerations, and recent progress. Advanced Materials 25(42):5997–6038.
3. Shinichiro O, Yosuke H, Lu J, Yohei Y, Genki A, Emi H et al. 2018. 58–1: Invited paper: High resolution IPS-LCDs fabricated with transparent polyimide substrates. SID Symposium Digest of Technical Papers 49(1): 764–767.
4. Murashige T, Fujikake H, Sato H, Kikuchi H, Kurita T, Sato F. 2004. Polymer wall formation using liquid-crystal/polymer phase separation induced on patterned polyimide films. Japanese Journal of Applied Physics 43(12B): L1578–L1580.
5. Mach P, Rodriguez SJ, Nortrup R, Wiltzius P, Rogers JA. 2001. Monolithically integrated, flexible display of polymer-dispersed liquid crystal driven by rubber-stamped organic thin-film transistors. Applied Physics Letters 78:3592–3594.
6. Zeng XY, Zhang QK, Yu RM, Lu CZ. 2010. A new transparent conductor: Silver nanowire film buried at the surface of a transparent polymer. Advanced Materials 22(40):4484–4488. DOI: 10.1002/adma.201001811.
7. Gaynor W, Hofmann S, Christoforo MG, Sachse C, Mehra S, Salleo A et al. 2013. Color in the corners: ITO-free white OLEDs with angular color stability. Advanced Materials 25(29):4006–4013. DOI: 10.1002/adma.201300923.
8. Lim S, Son D, Kim J, Lee YB, Song J-K, Choi S et al. 2015. Transparent and stretchable interactive human machine interface based on patterned graphene heterostructures. Advanced Functional Materials 25(3):375–383. DOI: 10.1002/adfm.201402987.
9. Hwang B-U, Lee J-H, Trung TQ, Roh E, Kim D-I, Kim S-W et al. Transparent stretchable self-powered patchable sensor platform with ultrasensitive recognition of human activities. ACS Nano. 2015;9(9):8801–8810. DOI: 10.1021/acsnano.5b01835.
10. Lee JH, Huynh-Nguyen B-C, Ko E, Kim JH, Seong GH. 2016. Fabrication of flexible, transparent silver nanowire electrodes for amperometric detection of hydrogen peroxide. Sensors and Actuators B: Chemical 224:789–797. DOI: 10.1016/j.snb.2015.11.006.

11. Hu W, Niu X, Zhao R, Pei Q. 2013. Elastomeric transparent capacitive sensors based on an interpenetrating composite of silver nanowires and polyurethane. Applied Physics Letters 102(8):083303. DOI: 10.1063/1.4794143.
12. Kim D-H, Yu K-C, Kim Y, Kim J-W. 2015. Highly stretchable and mechanically stable transparent electrode based on composite of silver nanowires and polyurethane–urea. ACS Applied Materials & Interfaces 7(28):15214–15222. DOI: 10.1021/acsami.5b04693.
13. Jeong CK, Lee J, Han S, Ryu J, Hwang GT, Park DY et al. 2015. A hyper-stretchable elastic composite energy harvester. Advanced Materials 27(18):2866–2875. DOI: 10.1002/ adma.201500367.
14. Ho X, Cheng CK, Tey JN, Wei J. 2015. Tunable strain gauges based on two-dimensional silver nanowire networks. Nanotechnology 26(19):195504. DOI: 10.1088/0957-4484/26/ 19/195504.
15. Kim H, Gilmore C, Pique A, Horwitz J, Mattoussi H, Murata H et al. 1999. Electrical, optical, and structural properties of indium–tin–oxide thin films for organic light-emitting devices. Journal of Applied Physics 86(11):6451–6461.
16. Charlier J-C, Blase X, Roche S. 2007. Electronic and transport properties of nanotubes. Reviews of Modern Physics 79:677.
17. Ma K, Yan X, Guo Y, Xiao Y. 2011. Electronic transport properties of junctions between carbon nanotubes and graphene nanoribbons. The European Physical Journal B: Condensed Matter and Complex Systems 83(4):487–492.
18. Hu L, Kim HS, Lee J-Y, Peumans P, Cui Y. 2010. Scalable coating and properties of transparent, flexible, silver nanowire electrodes. ACS Nano 4:2955–2963.
19. Shirota Y. 2000. Organic materials for electronic and optoelectronic devices. Journal of Materials Chemistry 10:1–25.
20. Oyama M. 2010. Recent nanoarchitectures in metal nanoparticle-modified electrodes for electroanalysis. Analytical Sciences 26(1):1–12.
21. Jiang X, Wong F, Fung M, Lee S. 2003. Aluminum-doped zinc oxide films as transparent conductive electrode for organic light-emitting devices. Applied Physics Letters 83(9):1875–1877.
22. Wang X, Zhi L, Müllen K. 2008. Transparent, Conductive Graphene Electrodes for Dye-Sensitized Solar Cells. Nano Letters 8(1):323–327.
23. Wu J, Becerril HA, Bao Z, Liu Z, Chen Y, Peumans P. 2008. Organic solar cells with solution-processed graphene transparent electrodes. Applied Physics Letters 92(26):263302.
24. Wang J, Fei F, Luo Q, Nie S, Wu N, Chen X et al. 2017. Modification of the highly conductive PEDOT:PSS layer for use in silver nanogrid electrodes for flexible inverted polymer solar cells. ACS Applied Materials & Interfaces 9(8):7834–7842. DOI: 10.1021/acsami.6b16341.
25. Zhu G, Wang H, Zhang L. 2016. A comparative study on electrosorption behavior of carbon nanotubes electrodes fabricated via different methods. Chemical Physics Letters. 649:15–18. DOI: 10.1016/j.cplett.2016.02.027.
26. Imazu N, Fujigaya T, Nakashima N. 2016. Fabrication of flexible transparent conductive films from long double-walled carbon nanotubes. Science and Technology of Advanced Materials 15(2):025005. DOI: 10.1088/1468–6996/15/2/025005.
27. Kobayashi T, Bando M, Kimura N, Shimizu K, Kadono K, Umezu N et al. 2013. Production of a 100-m-long high-quality graphene transparent conductive film by roll-to-roll chemical vapor deposition and transfer process. Applied Physics Letters 102(2):023112. DOI: 10.1063/1.4776707.

28. Zhang C, Zhu Y, Yi P, Peng L, Lai X. 2017. Fabrication of flexible silver nanowire conductive films and transmittance improvement based on moth-eye nanostructure array. Journal of Micromechanics & Microengineering 27(7):075010.
29. Lee JY, Connor ST, Cui Y, Peumans P. 2008. Solution-processed metal nanowire mesh transparent electrodes. Nano Letters 8(2):689–692. DOI: 10.1021/nl073296g.
30. Ghaffarzadeh K, Yamamoto Y, Zervos H. 2016. *Conductive Ink Markets 2016–2026: Forecasts, Technologies, Players*. Boston, MA: IDTechEx Ltd.
31. Lee J, Lee P, Lee HB, Hong S, Lee I, Yeo J et al. 2013. Room-temperature nanosoldering of a very long metal nanowire network by conducting-polymer-assisted joining for a flexible touch-panel application. Advanced Functional Materials 23(34):4171–4176. DOI: 10.1002/adfm.201203802.
32. Lee J, Lee P, Lee H, Lee D, Lee SS, Ko SH. 2012. Very long Ag nanowire synthesis and its application in a highly transparent, conductive and flexible metal electrode touch panel. Nanoscale 4(20):6408.
33. Kim Y, Song C-H, Kwak M-G, Ju B-K, Kim J-W. 2015. Flexible touch sensor with finely patterned ag nanowires buried at the surface of a colorless polyimide film. RSC Advances 5(53):42500–42505. DOI: 10.1039/c5ra01657f.
34. Mayousse C, Celle C, Moreau E, Mainguet JF, Carella A, Simonato JP. 2013. Improvements in purification of silver nanowires by decantation and fabrication of flexible transparent electrodes. Application to capacitive touch sensors. Nanotechnology 24(21):215501. DOI: 10.1088/0957–4484/24/21/215501.
35. Kim Y, Kim J-W. 2016. Silver nanowire networks embedded in urethane acrylate for flexible capacitive touch sensor. Applied Surface Science 363:1–6. DOI: 10.1016/j.apsusc. 2015.11.052.
36. Cann M, Large MJ, Henley SJ, Milne D, Sato T, Chan H et al. 2016. High performance transparent multi-touch sensors based on silver nanowires. Materials Today Communications 7:42–50. DOI: 10.1016/j.mtcomm.2016.03.005.
37. Choi K-H, Kim J, Noh Y-J, Na S-I, Kim H-K. 2013. Ag nanowire-embedded ITO films as a near-infrared transparent and flexible anode for flexible organic solar cells. Solar Energy Materials and Solar Cells 110:147–153. DOI: 10.1016/j.solmat.2012.12.022.
38. Guo F, Zhu X, Forberich K, Krantz J, Stubhan T, Salinas M et al. 2013. ITO-free and fully solution-processed semitransparent organic solar cells with high fill factors. Advanced Energy Materials 3(8):1062–1067. DOI: 10.1002/aenm.201300100.
39. Leem DS, Edwards A, Faist M, Nelson J, Bradley DD, de Mello JC. 2011. Efficient organic solar cells with solution-processed silver nanowire electrodes. Advanced Materials 23(38):4371–4375. DOI: 10.1002/adma.201100871.
40. Angmo D, Andersen TR, Bentzen JJ, Helgesen M, Søndergaard RR, Jørgensen M et al. 2015. Roll-to-roll printed silver nanowire semi transparent electrodes for fully ambient solution-processed tandem polymer solar cells. Advanced Functional Materials 25(28):4539–4547. DOI: 10.1002/adfm.201501887.
41. Karakawa M, Tokuno T, Nogi M, Aso Y, Suganuma K. 2017. Silver nanowire networks as a transparent printable electrode for organic photovoltaic cells. Electrochemistry 85(5):245–248.
42. Kang S, Cho S, Shanker R, Lee H, Park J, Um D-S et al. 2018. Transparent and conductive nanomembranes with orthogonal silver nanowire arrays for skin-attachable loudspeakers and microphones. Applied Science Engineers, Science Advances 4:eaas8772.
43. Chen Y-Y, Sun Y, Zhu Q-B, Bing-Wei W, Yan X, Qiu S et al. 2018. High-throughput fabrication of flexible and transparent all-carbon nanotube electronics. Advanced Science 5(5):1700965. DOI: 10.1002/advs.201700965.

44. Huang S, Liu Y, Zhao Y, Ren Z, Guo C. 2018. Flexible electronics: Stretchable electrodes and their future. Advanced Functional Materials 29(6): 1805924.
45. Kim KK, Hong S, Cho HM, Lee J, Suh YD, Ham D et al. 2015. Highly sensitive and stretchable multidimensional strain sensor with prestrained anisotropic metal nanowire percolation networks. Nano Letters 15(8):5240–5207.
46. Han S, Hong S, Ham J, Yeo J, Lee J, Kang B et al. 2014. Fast plasmonic laser nanowelding for a Cu-nanowire percolation network for flexible transparent conductors and stretchable electronics. Advanced Materials 26(33):5808–5814.

Index

Note: Page numbers in *italics* indicate figures and in **bold** indicate tables on the corresponding pages.

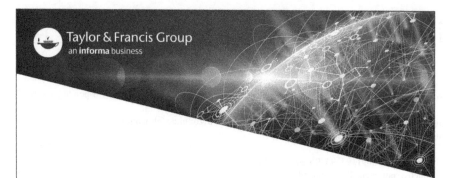